Undergraduate Programs and Mathematical Sciences: CUPM Curriculum Guide 2004

A report by the
Committee on the Undergraduate Program in Mathematics of
The Mathematical Association of America

This report is available at www.maa.org/cupm/.

CUPM Guide 2004 was prepared with support from the National Science Foundation and from the Calculus Consortium for Higher Education.

Library of Congress Catalog Card Number 2004100651

ISBN 0-88385-814-2

Printed in the United States of America

Current printing (last digit):
10 9 8 7 6 5 4 3 2

Undergraduate Programs and Courses in the Mathematical Sciences: CUPM Curriculum Guide 2004

A report by the
Committee on the Undergraduate Program in Mathematics of
The Mathematical Association of America

CUPM Writing Team
William Barker
David Bressoud
Susanna Epp
Susan Ganter
Bill Haver
Harriet Pollatsek (chair)

Published and Distributed by
The Mathematical Association of America

Committee on the Undergraduate Program in Mathematics (CUPM)

Dora Ahmadi, Morehead State University
Thomas Banchoff, Brown University
*William Barker, Bowdoin College
*Lynne Bauer, Carleton College
*Thomas Berger (Chair 1994–2000), Colby College
David Bressoud, Macalester College
*Amy Cohen, Rutgers University
Lynda Danielson, Albertson College
*Susanna Epp, DePaul University
*Naomi Fisher, University of Illinois at Chicago
*Joseph Gallian, University of Minnesota-Duluth
Ramesh Gangolli, University of Washington
*Frank Giordano, US Military Academy (ret) and COMAP
Jose Giraldo, Texas A&M University, Corpus Christi
*Bill Haver, Virginia Commonwealth University
Dianne Hermann, University of Chicago
*Peter Hinman, University of Michigan
*Herbert Kasube, Bradley University
Daniel Maki, Indiana University
Joseph Malkevitch, York College CUNY
Mercedes McGowen, Harper Community College
Harriet Pollatsek (Chair 2000–2003), Mount Holyoke College
Marilyn Repsher, Jacksonville University
Allan Rossman, California Polytechnic State University, San Luis Obispo
Kathleen Snook, US Military Academy (ret)
*Olaf Stackelberg, Kent State University
Michael Starbird, University of Texas

* Term on CUPM ended before completion of *CUPM Guide2004*

Curriculum Project Steering Committee

William Barker
Thomas Berger
David Bressoud
Susanna Epp
Susan Ganter
Bill Haver
Herbert Kasube
Harriet Pollatsek (chair)

Susan Ganter chairs the CUPM subcommittee on Curriculum Renewal Across the First Two Years (CRAFTY); William Barker is former chair of CRAFTY.

The writing team acknowledges with gratitude the assistance of Barry Cipra (preliminary drafting), Kathleen Snook (compiling and editing Illustrative Resources during the winter and spring of 2003), and Thomas Rishel and Michael Pearson (providing MAA staff support).

Contents

Executive Summary 1

Introduction 3

Part I: Recommendations for Departments, Programs, and all
Courses in the Mathematical Sciences 11

Part II: Additional Recommendations Concerning Specific Student Audiences 27

    A. Students taking general education or introductory collegiate courses
       in the mathematical sciences 27

    B. Students majoring in partner disciplines 32

    C. Students majoring in the mathematical sciences 42

    D. Mathematical sciences majors with specific career goals 52

Illustrative Resources 63

References 69

Appendices

    1. The CUPM Curriculum Initiative 73

    2. The Curriculum Foundations project 77

    3. Data on numbers of majors 81

    4. Data on student goals, department practices, advanced courses 85

    5. Summary of recommendations 91

    6. Sample questions for department self-study 97

Index 101

Executive Summary

The Mathematical Association of America's Committee on the Undergraduate Program in Mathematics (CUPM) is charged with making recommendations to guide mathematics departments in designing curricula for their undergraduate students. CUPM began issuing reports in 1953, updating them at roughly 10-year intervals. *Undergraduate Programs and Courses in the Mathematical Sciences: CUPM Curriculum Guide 2004* is based on four years of work,[1] including extensive consultation with mathematicians and members of partner disciplines.[2] Available at `www.maa.org/cupm/`, *CUPM Guide 2004* contains the recommendations unanimously approved by CUPM in January 2003.

Many recommendations in *CUPM Guide 2004* echo those in previous CUPM reports, but some are new. In particular, previous reports focused on the undergraduate program for mathematics majors, although with a steadily broadening definition of the major. *CUPM Guide 2004* addresses the *entire* college-level mathematics curriculum, for *all* students, even those who take just one course. *CUPM Guide 2004* is based on six fundamental recommendations for departments, programs and all courses in the mathematical sciences. The MAA Board of Governors approved these six recommendations at their Mathfest 2003 meeting.

Recommendation 1: *Mathematical sciences departments should*

- *Understand the strengths, weaknesses, career plans, fields of study, and aspirations of the students enrolled in mathematics courses;*

- *Determine the extent to which the goals of courses and programs offered are aligned with the needs of students as well as the extent to which these goals are achieved;*

- *Continually strengthen courses and programs to better align with student needs, and assess the effectiveness of such efforts.*

Recommendation 2: *Every course should incorporate activities that will help all students progress in developing analytical, critical reasoning, problem-solving, and communication skills and acquiring mathematical habits of mind. More specifically, these activities should be designed to advance and measure students' progress in learning to*

- *State problems carefully, modify problems when necessary to make them tractable, articulate assumptions, appreciate the value of precise definition, reason logically to conclusions, and interpret results intelligently;*

- *Approach problem solving with a willingness to try multiple approaches, persist in the face of difficulties, assess the correctness of solutions, explore examples, pose questions, and devise and test conjectures;*

---

[1] Supported by the National Science Foundation and the Calculus Consortium for Higher Education.

[2] Reports from a series of workshops on the mathematics curriculum with members of partner disciplines are contained in *The Curriculum Foundations Project: Voices of the Partner Disciplines,* edited and with an introduction by Susan Ganter and William Barker (MAA, 2004).

- *Read mathematics with understanding and communicate mathematical ideas with clarity and coherence through writing and speaking.*

Recommendation 3: *Every course should strive to*
- *Present key ideas and concepts from a variety of perspectives;*
- *Employ a broad range of examples and applications to motivate and illustrate the material;*
- *Promote awareness of connections to other subjects (both in and out of the mathematical sciences) and strengthen each student's ability to apply the course material to these subjects;*
- *Introduce contemporary topics from the mathematical sciences and their applications, and enhance student perceptions of the vitality and importance of mathematics in the modern world.*

Recommendation 4: *Mathematical sciences departments should encourage and support faculty collaboration with colleagues from other departments to modify and develop mathematics courses, create joint or cooperative majors, devise undergraduate research projects, and possibly team-teach courses or units within courses.*

Recommendation 5: *At every level of the curriculum, some courses should incorporate activities that will help all students progress in learning to use technology*
- *Appropriately and effectively as a tool for solving problems;*
- *As an aid to understanding mathematical ideas.*

Recommendation 6: *Mathematical sciences departments and institutional administrators should encourage, support and reward faculty efforts to improve the efficacy of teaching and strengthen curricula.*

Part I of *CUPM Guide 2004* elaborates on these recommendations and suggests ways that a department can evaluate its progress in meeting them. Part II contains supplementary recommendations concerning particular student audiences:

- A. Students taking general education or introductory courses in the mathematical sciences;
- B. Students majoring in partner disciplines, including those preparing to teach mathematics in elementary or middle school;
- C. Students majoring in the mathematical sciences;
- D. Mathematical sciences majors with specific career goals: secondary school teaching, entering the non-academic workforce, and preparing for post-baccalaureate study in the mathematical sciences and allied disciplines.

Specific methods for implementation are not prescribed, but the online document *Illustrative Resources for CUPM Guide 2004* at www.maa.org/cupm/ describes a variety of experiences and resources associated with these recommendations. These illustrative examples are not endorsed by CUPM, but they may serve as a starting point for departments considering enhancement of their programs. Pointers to additional resources, such as websites (with active links) and publications, are also given.

Introduction

Mathematics is universal: it underlies modern technology, informs public policy, plays an essential role in many disciplines, and enchants the mind. At the start of the twenty-first century, the undergraduate study of mathematics can and should be a vital and engaging part of preparation for many careers and for well-informed citizenship. In the *CUPM Guide 2004*, the term 'mathematics' is generally synonymous with 'mathematical sciences' and refers to a collection of mathematics-related disciplines, including, but not necessarily limited to, pure and applied mathematics, mathematics education, computational mathematics, operations research, and statistics. Departments of mathematical sciences can and should play a central role in their institutions' undergraduate programs. The *CUPM Guide 2004*[3] calls on mathematicians and mathematics departments to rethink the full range of their undergraduate curriculum and co-curriculum to ensure the best possible mathematical education for all their students, from liberal arts students taking just one course to students majoring in the mathematical sciences.

The need for action

Over the past one hundred years mathematics has become more important to more disciplines than ever before. At the same time both the number and diversity of students in post-secondary education and the variety of their mathematical backgrounds have increased dramatically. Additionally, computer technology has forever altered the way mathematics is used in the workplace; from retail store registers to financial institutions to laboratories doing advanced scientific research.

These developments present unprecedented curricular challenges to departments of mathematical sciences—challenges many departments and individual faculty members are engaged in meeting. During the past twenty years there has been an explosive increase in the number of presentations and publications on issues and innovations in the teaching of post-secondary mathematics.[4] This activity reflects a growing movement to address the undergraduate mathematics curriculum conscientiously and creatively.

Yet at many institutions, students and faculty outside mathematics perceive mathematics departments as uninterested in adapting instruction to new circumstances, especially to the needs of non-majors and those in entry-level courses. Indeed, many view the formal study of mathematics as irrelevant or tangential to the needs of today's society. They see mathematics departments as disconnected from other disciplines except through a service component that they believe is accepted only reluctantly and executed without inspiration or effectiveness. Such views were expressed by a majority of the academic deans at the

---

[3] The MAA publication *Guidelines for Programs and Departments in Undergraduate Mathematical Sciences*, MAA, 2001, available at `www.maa.org/guidelines/guidelines.html`, complements the *CUPM Guide 2004* and other curricular reports by presenting a set of recommendations that deal with a broad range of structural issues that face mathematical sciences departments and their institutional administrations.

[4] For example, at the January 2003 Joint Mathematics Meetings, more than one third of the talks concerned mathematics education.

research universities sampled for the American Mathematical Society (AMS) study *Towards Excellence*: "The prevalent theme in every discussion [with deans] was the insularity of mathematics. Mathematicians do not interact with other departments or with faculty outside mathematics, many deans claimed. ... The deans ... seemed to view mathematics departments as excessively inward looking."[5]

This perception is often due more to poor communication than to a lack of effort or good intention. At the least, it points to the need for better communication. But there are other indicators that all is not well. National data provide clear evidence that undergraduate mathematics programs are under serious pressure, with decreasing numbers of mathematics majors and declining enrollment in advanced mathematics courses.[6] From 1985 to 2000 the total number of bachelor's degrees awarded annually in the U.S. rose 25% and the number of science and technology degrees grew 20%. However, data collected by the Conference Board of the Mathematical Sciences (CBMS)[7] show that the total number of degrees awarded annually by mathematics and statistics departments, including those in secondary mathematics education, stayed essentially flat during this 15-year period. In fact, the annual total in these departments fell 4% between 1995 and 2000, and the number of annual degrees in mathematics fell 19% in the 1990s. The drop in mathematics degrees occurred at the same time as the need for new teachers of secondary mathematics grew more acute.

One might expect the increase in science and technology degrees to translate into higher enrollment in advanced mathematics courses as allied subjects. In fact, the opposite has occurred: enrollment in advanced courses taught in mathematics departments has fallen, dropping 25% from 1985 to 2000; an increase from 1995 to 2000, while encouraging, has not returned enrollment to 1990 levels. Further, CBMS data show that even the *availability* of advanced courses has declined in the past five years, as the percentages of departments offering several typical courses[8] has decreased, in some cases by more than 20%. This trend is unfortunate, not only for the health of the mathematical sciences major but also because the health of disciplines that use mathematics—and by extension the health of society—is enhanced when a significant number of students are knowledgeable about the advanced mathematics that is relevant to their fields.

Striking successes at a number of colleges and universities demonstrate that these perceptions can be changed and these trends reversed. For instance, the MAA volume *Models That Work: Case Studies in Effective Undergraduate Mathematics Programs,*[9] summarizes effective practices at a set of mathematics departments that have excelled in (i) attracting and training large numbers of mathematics majors, or (ii) preparing students to pursue advanced study in mathematics, or (iii) preparing future school mathematics teachers, or (iv) attracting and training underrepresented groups in mathematics. Site-visits to ten departments and information on a number of others revealed "no single key to a successful undergraduate program in mathematics." However, there were common features. "What was a bit unexpected was the common attitude in effective programs that the faculty are not satisfied with the current program. They are constantly trying innovations and looking for improvement."

---

[5] *Towards Excellence: Leading a Doctoral Mathematics Department in the 21st Century*, American Mathematical Society Task Force on Excellence, J. Ewing editor, AMS, 1999, p. 65.

[6] See Appendix 3 for further analysis of data on numbers of majors and the supply of secondary teachers of mathematics, and Appendix 4 for data on enrollment in and availability of advanced courses.

[7] *CBMS 2000: Statistical Abstract of Undergraduate Programs in the Mathematical Sciences in the United States*, D. Lutzer, J. Maxwell and S. Rodi, AMS, 2002.

[8] Including algebra, analysis, geometry, mathematical modeling and applied mathematics; see Table 4-3 in Appendix 4.

[9] MAA Notes **38** (1995), Alan C. Tucker, editor.

Areas for attention and action

Mathematics departments need to serve *all* students well—not only those who major in the mathematical or physical sciences. The following steps will help departments reach this goal.

- Design undergraduate programs to address the broad array of problems in the diverse disciplines that are making increasing use of mathematics.

- Guide students to learn mathematics in a way that helps them to better understand its place in society: its meaning, its history, and its uses. Such understanding is often lacking even among students who major in mathematics.

- Employ a broad range of instructional techniques, and require students to confront, explore, and communicate important ideas of modern mathematics and the uses of mathematics in society. Students need more classroom experiences in which they learn to think, to do, to analyze—not just to memorize and reproduce theories or algorithms.

- Understand and respond to the impact of computer technology on course content and instructional techniques.

- Encourage and support faculty in this work—a task both for departments and for administrations.

The *CUPM Guide 2004* presents six general recommendations to assist mathematics departments in the design and teaching of all of their courses and programs. It also contains supplementary recommendations for particular student audiences.

1. Understand the student population and evaluate courses and programs

In summarizing the common features of the programs described in *Models That Work,* the authors wrote that one of the "states of mind that underlie faculty attitudes in effective programs" is "teaching for the students one has, not the students one wished one had." *Towards Excellence* echoes this theme: "Mathematics departments should position themselves to receive new or reallocated resources by meeting the needs of their institutions. That does not mean sacrificing the intellectual integrity of an academic program, nor does it mean relegating mathematics to a mere service role. It *does* mean fulfilling a bargain with the institution in which one lives, and for most departments a major part of that bargain involves instruction."[10]

Recommendation 1: Mathematical sciences departments should

- Understand the strengths, weaknesses, career plans, fields of study, and aspirations of the students enrolled in mathematics courses;

- Determine the extent to which the goals of courses and programs offered are aligned with the needs of students as well as the extent to which these goals are achieved;

- Continually strengthen courses and programs to better align with student needs, and assess the effectiveness of such efforts.

2. Develop mathematical thinking and communication skills

The power of mathematical thinking—pattern recognition, generalization, abstraction, problem solving, careful analysis, and rigorous argument—is important for every citizen. It is highly valued by employers and by other disciplines but widely misunderstood and undervalued by students.

---

[10] *Towards Excellence*, p. xiii.

Communication is integral to learning and using mathematics, and skill in communicating is commonly listed as the most important quality employers seek in a prospective employee.[11] However, many students expect mathematics classes to be wordless islands where they won't be asked to read, write, or discuss ideas.

Appropriate instructional approaches to reasoning and proof have been passionately debated among mathematicians for decades, but with a greater sense of urgency during the last twenty years. While much remains to be learned about how best to teach reasoning and proof skills—as well as how best to improve communication skills—a variety of strategies can help students progress. Mathematics faculty should deliver an unambiguous message concerning the importance of mathematical reasoning and communication skills and adopt instructional methods and curriculum content that develop these skills. Designing a curriculum that develops these skills effectively and at appropriate levels for all students is one of the biggest and most important challenges for mathematics departments.

Recommendation 2: Every course should incorporate activities that will help all students progress in developing analytical, critical reasoning, problem-solving, and communication skills and acquiring mathematical habits of mind. More specifically, these activities should be designed to advance and measure students' progress in learning to

- State problems carefully, modify problems when necessary to make them tractable, articulate assumptions, appreciate the value of precise definition, reason logically to conclusions, and interpret results intelligently;

- Approach problem solving with a willingness to try multiple approaches, persist in the face of difficulties, assess the correctness of solutions, explore examples, pose questions, and devise and test conjectures;

- Read mathematics with understanding and communicate mathematical ideas with clarity and coherence through writing and speaking.

3. Communicate the breadth and interconnections of the mathematical sciences

Many students do not see the connections between mathematics and other disciplines or between mathematics and the world in which they live. Too often they leave mathematics courses with a superficial mastery of skills that they are unable to apply in non-routine settings and whose importance to their future careers is unrecognized. Conceptual understanding of mathematical ideas and facility in mathematical thinking are essential for both applications and further study of mathematics, yet they are often lost in a long list of required topics and computational techniques. Even when students successfully apply mathematical techniques to problems, they are often unable to interpret their results effectively or communicate them with clarity.

The beauty, creativity, and intellectual power of mathematics and its contemporary challenges and discoveries, are often unknown and unappreciated. The interplay between differing perspectives—continuous and discrete, deterministic and stochastic, algebraic and geometric, exact and approximate—is appreciated by very few students, even though flexible use of these varying perspectives is critical for applications and for learning new mathematics.

Recommendation 3: Every course should strive to

- Present key ideas and concepts from a variety of perspectives;

- Employ a broad range of examples and applications to motivate and illustrate the material;

---

[11] See, for instance, surveys by the National Association of Colleges and Employers, www.naceweb.org.

- Promote awareness of connections to other subjects (both in and out of the mathematical sciences) and strengthen each student's ability to apply the course material to these subjects;
- Introduce contemporary topics from the mathematical sciences and their applications, and enhance student perceptions of the vitality and importance of mathematics in the modern world.

4. Promote interdisciplinary cooperation

Mathematics programs have traditionally drawn heavily from the physical sciences for applications. In recent years, mathematics has come to play a significant role in far more disciplines, but many mathematics programs have not adjusted to this new reality.

Mathematics departments should seize the opportunity to harness the growing awareness in other disciplines of the power and importance of mathematical methods. A curriculum developed in consultation with other disciplines that includes a variety of courses and degree options can attract more students, help them learn important mathematical ideas, retain more students for intermediate and advanced coursework, strengthen their ability to apply mathematics to other areas, and improve the quantity and quality of the mathematics majors and minors.

Recommendation 4: Mathematical sciences departments should encourage and support faculty collaboration with colleagues from other departments to modify and develop mathematics courses, create joint or cooperative majors, devise undergraduate research projects, and possibly team-teach courses or units within courses.

5. Use computer technology to support problem solving and to promote understanding

Recent advances in desktop and handheld computer technology can be used to improve the pedagogy and content of mathematics courses at all levels. Some mathematical ideas and procedures have become less important because of these emerging technological tools; others have gained importance. The 2001 *MAA Guidelines for Programs and Departments in Undergraduate Mathematical Sciences* recommended that departments "should employ technology in ways that foster teaching and learning, increase the students' understanding of mathematical concepts, and prepare students for the use of technology in their careers or in their graduate study." [12] However, much remains to be done to effectively meet the challenges posed by the growth of technology.

Recommendation 5: At every level of the curriculum, some courses should incorporate activities that will help all students progress in learning to use technology

- Appropriately and effectively as a tool for solving problems;
- As an aid to understanding mathematical ideas.

6. Provide faculty support for curricular and instructional improvement

Many of the recommendations in the *CUPM Guide 2004*, including collaborating with colleagues in other disciplines, adapting material from other parts of mathematics or from other disciplines for use in teaching, evaluating student writing, and making effective use of technology, require time and effort from faculty beyond what they might ordinarily devote to the revision and creation of courses. Departments and administrators need to acknowledge that meeting these recommendations makes substantial demands on faculty (and, in some cases, on graduate teaching assistants and other temporary or part-time instructors).

---

[12]*Guidelines for Programs and Departments in Undergraduate Mathematical Sciences*, MAA, 2001.

Recommendation 6: Mathematical sciences departments and institutional administrators should encourage, support and reward faculty efforts to improve the efficacy of teaching and strengthen curricula.

Using the CUPM Guide 2004

Part I of the *CUPM Guide 2004* elaborates on and specifies the meaning of the six general recommendations as well as suggesting ways that a department can evaluate progress in meeting them. Part II contains supplementary recommendations concerning particular student audiences.

Some students major in fields that do not require specific mathematical preparation. They may take one course in mathematics, perhaps to satisfy a general education requirement of their institution or major program. Section A of Part II addresses the needs of these students, many of whom—especially among the hundreds of thousands enrolled each semester in courses called College Algebra—are not optimally served by the mathematics courses they take.

Partner disciplines are those whose majors are required to take one or more specific mathematics courses. These disciplines vary by institution but usually include the physical sciences, the life sciences, computer science, engineering, economics, business, education, and often several social sciences. Recommendations concerning these students are in Section B of Part II, including those for pre-service K–8 teachers.

Section C of Part II contains recommendations concerning students majoring in the mathematical sciences. The recommendations urge departments to learn the probable career paths and needs of their majors and offer them a flexible program that provides appropriate breadth and depth. Section D contains further recommendations for mathematical sciences majors preparing to teach secondary school mathematics, planning for non-academic employment, or intending post-baccalaureate study.

There are many ways to carry out each recommendation, and different choices will be appropriate in different institutional settings. Consequently, these recommendations rarely specify particular courses or syllabi. That doesn't mean "anything goes." Indeed, each recommendation is accompanied by measures to help a department gauge its effectiveness.[13] As stated in the *MAA Guidelines for Programs and Departments in Undergraduate Mathematical Sciences*, "These measures will, of necessity, be multidimensional since no single statistic can adequately represent departmental performance with respect to most departmental goals. Measures of student learning and other student outcomes should be included."[14] Course syllabi and sample assignments, along with their contribution to students' grades, are other valuable measures.

Although no specific methods for implementation are outlined, the hyper-linked web document *Illustrative Resources for CUPM Guide 2004* at `www.maa.org/cupm/` is designed to help departments implement and improve practices to satisfy the recommendations. It is organized and numbered the same way as the recommendations in Parts I and II. A variety of examples, including assignments, courses (with suggested syllabi and texts), and programs are provided for each recommendation. The examples range along a continuum, from modest first steps and small changes that can be easily effected to more ambitious efforts. Pointers to additional resources, such as websites and publications, are also given, with live links where appropriate.

These recommendations have been reduced to a core judged essential for building and supporting department strength and effectively meeting department obligations. They are not a wish list for an ideal future department. Indeed, the reality is that departments at many institutions are coping with diminished

---

[13] Also see Appendix 6, Sample questions for department self-study.

[14] MAA 2001, available at `www.maa.org/guidelines/guidelines.html`.

human and financial resources and conflicting and escalating demands on faculty time. Moreover, meaningful change is never easy. Nonetheless, the use of the word "should" in a recommendation means that departments are expected to make a conscientious effort to achieve steady improvement until they are able to satisfy it.

Background for the recommendations

The Mathematical Association of America's Committee on the Undergraduate Program in Mathematics (CUPM) is charged with making recommendations to guide mathematics departments in designing curricula for their undergraduate students. CUPM began issuing reports in 1953, updating them at roughly 10-year intervals. In 1999 work began on the current recommendations. CUPM solicited position papers from prominent mathematicians and conducted panel discussions and focus groups at national meetings to obtain reactions to preliminary drafts of these recommendations. There has been extensive consultation with other professional societies in the mathematical sciences. From 1999 to 2002 CUPM's subcommittee on Curriculum Renewal Across the First Two Years (CRAFTY) conducted a series of workshops on the mathematics curriculum with participants from a broad range of partner disciplines.[15]

The MAA Board of Governors has endorsed the six fundamental recommendations for the design and teaching of all courses and programs. CUPM unanimously approved the supplementary recommendations in Part II of the *CUPM Guide 2004*. Many of the current recommendations echo those in previous CUPM reports, but some are new. In particular, previous reports focused on the undergraduate program for mathematics majors, although with a steadily broadening definition of the major in the 1981 and 1991 reports. The *CUPM Guide 2004*, in contrast, addresses the *entire* college-level mathematics curriculum, for *all* students, even those who take just one course.[16]

---

[15] See Appendices 1 and 2 for detailed accounts of CUPM's activities and of the CRAFTY workshops collectively known as the Curriculum Foundations project. The results of the project are contained in the MAA publication *The Curriculum Foundations Project: Voices of the Partner Disciplines,* edited and with an introduction "A Collective Vision: Voices of the Partner Disciplines" by Susan Ganter and William Barker (MAA, 2004).

[16] While attempting to address the college-level curriculum in mathematics more comprehensively, the *CUPM Guide 2004* does not discuss a number of important issues, including non-credit or developmental courses and articulation between institutions.

Part I. Recommendations for Departments, Programs, and all Courses in the Mathematical Sciences

1. Understand the student population and evaluate courses and programs

Mathematical sciences departments should

- *Understand the strengths, weaknesses, career plans, fields of study, and aspirations of the students enrolled in mathematics courses;*

- *Determine the extent to which the goals of courses and programs offered are aligned with the needs of students, as well as the extent to which these goals are achieved;*

- *Continually strengthen courses and programs to better align with student needs, and assess the effectiveness of such efforts.*

Higher education has changed dramatically since CUPM first convened in 1953. In that era barely 12% of secondary school graduates went on to college, and only about 60% of the eligible cohort were even getting through secondary school. Those who did go to college typically went as full-time students, completing a degree in the traditional four years.

Today, approximately 85% of the eligible cohort complete secondary school, and 25% of secondary school graduates complete at least four years of college.[17] Part-time study is common, especially at state universities. Two-year colleges enroll 44% of the undergraduates in the U.S. and 49% of those undergraduates who identify themselves as members of racial or ethnic minorities.[18] The college population also is more heterogeneous in ways that were inconceivable in the 1950s. Students bring a wide range of mathematical backgrounds to college. Some come with a full year of Advanced Placement calculus or International Baccalaureate mathematics. Many more enter with algebra as their most advanced mathematics subject.

Understand student backgrounds and needs. These changing demographics increase the responsibility of mathematics departments to understand the strengths, weaknesses, career plans, fields of study, and aspirations of students enrolled in mathematics courses. Given the large number of students enrolled, gaining this understanding is a major undertaking. However, it is essential. For example, information about what

---

[17] See www.ed.gov/pubs/YouthIndicators, Indicators 23 (School Enrollment) and 26 (School Completion).

[18] See "First Steps: The Role of the Two-Year College in the Preparation of Mathematics-Intensive Majors," by Susan S. Wood, in *CUPM Discussion Papers about Mathematics and the Mathematical Sciences in 2010: What Should Students Know?*, MAA Report, 2001. Also see *Crossroads in Mathematics: Standards for Introductory College Mathematics Before Calculus*, American Mathematical Association of Two-Year Colleges, 1995.

mathematics majors do after graduation is critical when designing programs for majors. What percentage go on to graduate programs in mathematics? to graduate programs in other fields? What fraction goes directly into the non-academic job market? Similarly, if students enrolled in a college algebra course will not need the technical computational skills typically emphasized in such a course, they should instead take a course focusing on concepts, applications and general mathematical thinking.

Many mathematics departments use a placement test to sort students into entry-level courses. Such a test may provide useful information on mathematical preparation, but it gives little or no information about the test takers' actual mathematical needs or academic interests.

One way to determine student needs in their major programs and desired careers is to consult with colleagues in other disciplines. Find out *what they think the mathematics department is teaching* and *what they think the department should be teaching.* CUPM has started this process through a series of workshops known as the Curriculum Foundations project, with partner disciplines ranging from information technology to mechanical engineering (see Appendix 2). The results are eye opening. It is a misimpression to think that colleagues in these disciplines only want students who can evaluate a logarithm and solve a differential or quadratic equation. Of course faculty in other disciplines want students to possess the computational skills required for their subjects. But they especially want students to possess conceptual understanding of the required mathematics, to have some experience with mathematical modeling, and to have the communication skills necessary for explaining the methods used to solve problems and the meaning of the solutions. The Curriculum Foundations reports can provide excellent starting points for such discussions on local campuses.

Determine if department programs serve student needs. The information gathered can serve as the basis for evaluating how well the goals of department courses and programs are aligned with student needs and the extent to which these goals are achieved. These efforts should encompass much more than the traditional classroom-level monitoring of student learning; department-level effectiveness of programs and personnel must be evaluated as well. Departments should regularly review course offerings and the topics within these courses, as well as the methods being used to teach and assess them.

The process of aligning the curriculum with student needs should be informed by sources outside of the mathematics department as well as within it. As faculty consult with colleagues in other disciplines about content, they also should discuss means of assessing how well the mathematics curriculum serves the needs of partner disciplines. Mathematics departments should further consult with alumni, employers and graduate programs at other institutions to obtain their impressions of the effectiveness of the mathematics curriculum and how the department could improve its program.

This may sound like a call for constant monitoring and micromanagement, but in fact it can be simple and direct. Faculty can collect much of the necessary information through simple questionnaires and informal conversations (as demonstrated by the examples in Illustrative Resources). Gradual modifications can be made to course syllabi, and information can be informally disseminated at department meetings and discussions over lunch. Faculty mentors can assist less experienced colleagues by initiating reciprocal classroom visits and taking time outside of class to talk about teaching ideas. Textbook adoption committees can use the assignment to revisit assumptions about courses.

In her essay on accountability in mathematics, Sandra Keith calls attention to a number of techniques to make the process non-disruptive and manageable, including starting simply using information already available, looking for "good news" to report and subsequently upgrading the relevant data, and tracking student achievement in courses.[19]

---

[19]"Accountability in Mathematics: Elevate the Objectives!" by S. Keith, in *CUPM Discussion papers about Mathematics and the Mathematical Sciences in 2010: What Should Students Know?* MAA Reports, 2001, p. 61.

Strengthen alignment and assess effectiveness. Strengthening programs to better align with student needs requires careful assessment of the success of such efforts. In 1995 CUPM endorsed a set of guidelines for establishing a cycle of assessment aimed at program improvement. They include the recommendation that mathematics departments ask three questions:

- What should students learn?
- How well are they learning?
- What should departments change so that future students will learn more and understand it better?

These questions provide a foundation for thinking about assessment. *Assessment Practices in Undergraduate Mathematics* (Gold *et al.*, 1999) contains over seventy case studies of assessment in mathematical sciences departments across the U.S. It provides a rich resource of examples to illustrate how assessment can be achieved in practice.[20] With support from the National Science Foundation (NSF), MAA is conducting a series of faculty development workshops in the area of assessment. The three-year project, entitled "Supporting Assessment in Undergraduate Mathematics" (SAUM), includes a workshop series, a volume of case studies and syntheses of case studies on assessment, and an informational website (`www.maa.org/saum/`). Further information about SAUM and other assessment resources developed by NSF and the National Academy of Science is included in Illustrative Resources.

2. Develop mathematical thinking and communication skills

Every course should incorporate activities that will help all students progress in developing analytical, critical reasoning, problem-solving, and communication skills_and acquiring mathematical habits of mind. More specifically, these activities should be designed to advance and measure students' progress in learning to

- *State problems carefully, modify problems when necessary to make them tractable, articulate assumptions, appreciate the value of precise definition, reason logically to conclusions, and interpret results intelligently;*
- *Approach problem solving with a willingness to try multiple approaches, persist in the face of difficulties, assess the correctness of solutions, explore examples, pose questions, and devise and test conjectures;*
- *Read mathematics with understanding and communicate mathematical ideas with clarity and coherence through writing and speaking.*

Terms like "analytical thinking" or "mathematical reasoning" are often used to describe those habits of mind that make employers want to hire mathematics majors, that lead departments and colleges to require mathematics courses, and that entice bright people into the further study of mathematics. Indeed, a striking finding from the Curriculum Foundations workshops[21] is that, contrary to the belief of many mathematicians, colleagues in partner disciplines (e.g., engineers, economists, and natural and computer scientists) value the precise, logical thinking that they perceive to be an integral part of mathematics and would like more emphasis on it in early collegiate mathematics instruction. Here are several such statements from the Curriculum Foundations reports:

- *Biotechnology and Environmental Technology*: "Requiring students to write an explanation of how they arrived at the result/answer and how they interpreted their results should reinforce writing skills and deductive reasoning."

---

[20] The entire volume, along with the MAA Assessment Guidelines, is available online at `www.maa.org/saum/`

[21] See Appendix 2 for a description of the Curriculum Foundations project.

- *Business*: "Courses should stress problem solving, with the incumbent recognition of ambiguities... [and] conceptual understanding (motivating the mathematics with the 'whys'—not just the 'hows')."

- *Chemistry*: "Logical, organized thinking and abstract reasoning are skills developed in mathematics courses that are essential for chemistry. At the physical chemistry level students must be able to follow logical reasoning and proofs, which is enabled by previous experience in mathematics courses."

- *Computer Science*: "Students should be comfortable with abstract thinking... they should have some facility with formal proofs, especially induction proofs."

- *Health and Life Sciences*: "Basic concepts that should be mastered include:... logic and mathematical thinking, generalization, deductive reasoning. At the conceptual level, students should be able to explain ... concepts in words."

- *Physics*: "The ability to actively think is the most important thing students need to get from mathematics education.... students should know that being able to integrate is quite different from understanding what integration is.... They must go beyond 'learning rules' to develop understanding."

Participants at the Curriculum Foundations workshop on the preparation of prospective mathematics majors also expressed this conviction in several parts of their report:

> The most important task of the first two years is to move students from a procedural/computational understanding of mathematics to a broad understanding encompassing logical reasoning, generalization, abstraction, and formal proof. The sooner this can be achieved, the better.... There should be an attempt to phase in logical language starting in the freshman year, rather than a sudden jump in the sophomore or junior year.... we should recognize our students' great variety of individual, cultural, and educational backgrounds. Students come with vastly different experience, skills, and learning styles. Introductory courses should provide experiences flexible enough to allow every student the opportunity both to reinforce existing strengths and to fill gaps.

Mathematical reasoning and problem-solving skills take a long time to develop and improvement is incremental. This has contributed to a growing recognition among mathematics departments of the need to consciously help students improve their reasoning and problem-solving skills. One result has been the increasing availability of transition-to-higher-mathematics courses and freshman-level discrete mathematics courses with a focus on logical argument and writing simple proofs.[22] The calculus renewal movement also has emphasized mathematical reasoning and problem solving, and modern statistics courses have been evolving along similar lines. Efforts are now underway to develop pre-calculus and other introductory mathematics courses according to the same principles.

Focusing on the development of reasoning skills from the earliest courses will

- improve the overall understanding of the nature and importance of mathematics, and

- elevate the level of mathematical competence in society.

Such a shift in focus also serves the self-interest of mathematical sciences departments because success will likely

- increase the pool of students capable of succeeding in higher-level mathematics courses, and

- encourage more of them to choose to enroll in higher-level mathematics courses.

---

[22] It is worth noting that much of the impetus for discrete mathematics courses has come from computer science departments, which have valued the contribution of these courses to students' general intellectual development as much—or more—than the specific topics they contain.

Learn to apply precise, logical reasoning to problem solving. Many students enter college with underdeveloped reasoning skills.[23] When a syllabus requires many topics to be covered in a short time and students' analytical skills are inadequate, emphasis on "problem solving" can easily turn into showing students how to imitate template solutions without real understanding. Even in linear algebra courses, sometimes a first introduction to definition and proof, pressure to teach computational techniques may override the goal of leading students to understand fundamental mathematical relationships.

Problem solving does not mean following recipes—it requires the application of careful, organized, creative thought to the analysis of complex and often ill-defined questions in a diverse array of situations. Problem solving is thus intimately connected with fundamental mathematical reasoning. Throughout any complex problem-solving process, bits of deductive logic, small proofs, and counterexamples are intermixed with visualization, flashes of intuition, and memory retrieval. It is thus artificial and ill advised to separate instruction in problem solving from other kinds of mathematical thinking.

An instructor can teach logical reasoning by devising classroom activities and assignments that focus on thinking skills. For example, an instructor can introduce a few examples of everyday "if-then" statements with the same logical interpretation as logical if-then statements and then ask the students to produce similar examples on their own. From this exercise students should observe the analogy between the mathematical logic and the everyday statements they have been considering. Or, if a student tries to justify a fact with a few examples, an instructor can give various properties that are sometimes true and sometimes false and have students find instances of both, illustrating the invalidity of "proof by example."

It takes time and effort to help students become better thinkers. Faculty must therefore seek a balance between expanding content and the need to help students improve their analytical, reasoning, and problem-solving skills. Faculty should realize that, while many students enter college with underdeveloped reasoning skills, the effort to help them improve is important. Even modest success makes a difference in students' lives, and when faculty approach the task with courtesy and respect, students appreciate their efforts.

Develop persistence and skill in exploration, conjecture, and generalization. Problem solving requires more than just solid mathematical reasoning — there are broad strategies and mental attitudes that students must identify, master, and internalize. To be successful problem solvers, students must learn persistence in the face of repeated rebuffs and flexibility in the choice of solution strategies. They must replace the question "did I get the right answer?" by "does my solution make sense?" Students must also learn to explore examples and special cases, to let new knowledge lead to new questions, to generalize and pose conjectures, yet to test all conjectures and retain a healthy skepticism toward unproven claims.

Instructors can stimulate students to generate questions and comments in response to readings, exercises, and presentations by modeling good questioning behavior: What do the words mean? What are some non-trivial examples? What motivates the material? What assumptions are being made? How do I know this is right? Through careful choice of problems and dialogue with students, faculty also can lead students to develop a more skeptical stance toward assertions: Does this make sense? Have all assumptions been enunciated?

Students need to be exposed to multi-stage projects that are built on exploration and conjecture and require persistence and flexibility for their solutions. More generally, at least some courses should be restructured to shift the burden from instructor to students for discovering and justifying results. A mathematical modeling course is an especially apt setting in which to make this shift and to raise students' awareness of the need to state problems carefully, articulate assumptions, and apply the mental and strate-

---

[23] See Illustrative Resources for some of the evidence of these difficulties.

gic tools of effective problem solving. Indeed, the value of modeling lies at least as much in the artful and creative thinking and thoughtful interpretation that it requires as in the connections it makes between mathematics and other disciplines.

Read and communicate mathematics with understanding and clarity. When employers are asked to list the qualities they seek in those who work for them, communication skills are invariably put first.[24] Mathematics faculty have an important share of the responsibility to help students improve their ability to communicate about technical matters. Most importantly, requiring students to read, write, and speak about mathematics helps them learn while providing instructors with valuable insight about levels of student understanding.

Some mathematics courses inadvertently encourage poor habits in reading, writing, and speaking. Instructors often do not require students to read the textbook, giving lectures that essentially reiterate it. In such situations, students mostly just scan the textbook for templates relevant to assigned homework problems. In fact, it is clear that many students don't know how to read mathematics. Students don't learn to read mathematics just because faculty ask them to do so. For these reasons instructors in most courses will need to continue to present material that is also in the text. But many instructors are demonstrating that it is possible gradually to teach students to read mathematics (see Illustrative Resources for examples).

Classroom discussion and informal oral presentations are important (but often overlooked) ways to help students improve thinking and problem-solving skills. For example, students can write homework solutions or partial solutions on the board each day at the beginning of class. Lively class discussions can arise from the students' board work, and they can set the tone for the rest of the class period. Many instructors develop problem solutions or proofs of simple statements by working in concert with their students, calling on individual students to explain what needs to be done next. In this way, students build confidence in their problem-solving abilities, while the instructor monitors the general level of understanding in the class.

Group work can further encourage students to verbalize mathematics both in and out of the classroom. Three MAA publications[25] describe the nuts and bolts of organizing cooperative work in a mathematics class. Rogers et al. (2001) includes the results of a faculty survey about the use of groups. Seventy-nine of the 94 faculty members surveyed reported that their students reacted "positively" or "very positively" to cooperative learning.

One of the most significant changes in mathematical pedagogy over the past couple of decades has been the increasing use of writing as a pedagogical tool. Requiring students to write about mathematics helps them learn and gives instructors valuable insight into the nature of their understanding. Some mathematics faculty are reluctant to require writing in the belief that they do not have the training to evaluate students' work. However, as many mathematics faculty are now demonstrating, grading and commenting on certain kinds of writing assignments does not require special training. For example, students can be given mathematical questions and asked to provide short written responses; the product is writing that all mathematicians can reasonably assess. Simply requiring answers in complete sentences can be a first step toward helping students better communicate about mathematical topics. Teaching them to write "From *A*, I know *B* because of *C*" can promote real growth in mathematical thinking. At more advanced levels, mathematics majors can be required to write substantial papers and present their results to faculty and students.

---

[24] See for example the websites www.careers-in-business.com, www.siam.org/mii and www.naceweb.org.

[25] *A Practical Guide to Cooperative Learning in Collegiate Mathematics.* Nancy L Hagelgans et al. eds. Mathematical Association of America, 1995; *Readings in Cooperative Learning for Undergraduate Mathematics.* Ed Dubinsky, David Mathews, and Barbara E. Reynolds, eds., Mathematical Association of America, 1997; and, *Cooperative Learning in Undergraduate Mathematics.* Elizabeth C. Rogers et al. eds. Mathematical Association of America, 2001.

The time needed to read and evaluate student writing can be reduced by a variety of techniques: cutting back on the number of assignments, asking for just a few sentences rather than lengthy essays, or having students write group reports. Some institutions make use of graduate or even undergraduate assistants to help with writing instruction and evaluation. Sessions at national and regional MAA meetings provide opportunities for faculty to learn additional ways to make student writing and presentations both beneficial and efficient. Examples of both modest and ambitious writing assignments can be found in Illustrative Resources, along with information about other useful resources.

3. Communicate the breadth and interconnections of the mathematical sciences

Every course should strive to

- *Present key ideas and concepts from a variety of perspectives;*

- *Employ a broad range of examples and applications to motivate and illustrate the material;*

- *Promote awareness of connections to other subjects (both in and out of the mathematical sciences) and strengthen each student's ability to apply the course material to these subjects;*

- *Introduce contemporary topics from the mathematical sciences and their applications, and enhance student perceptions of the vitality and importance of mathematics in the modern world.*

Mathematics is built from a rich variety of topics, perspectives, and applications. It includes the study of number systems, limits and calculus, algebraic functions and geometric transformations, deterministic and stochastic processes, topological shapes and geometric spaces, the separating of pattern from noise in real-world data, and the nature of logic itself. Great leaps of intuition are balanced by precise logical arguments. Practical problem solutions and the careful analysis of data are supported by powerful theoretical structures. Mathematics is inspired by the problems inherent in understanding the world, and it is strengthened and renewed by finding new ways to frame and answer questions about this world, both the physical world and the world of the intellect. Every mathematics department, in each of its courses, has a responsibility to communicate this mathematical richness, power, and beauty.

Within individual courses instructors should seek opportunities to present a broad range of perspectives on the material. It is impossible to incorporate every single viewpoint within every single course, but for every institution it is possible to include a broadly representative range of mathematical ideas and viewpoints within the full set of undergraduate courses. Departments should offer opportunities for students at various levels of mathematical sophistication to explore analytic, algebraic, geometric, discrete, statistical, and applied questions at a depth that is both accessible and challenging to them. And departments should ensure that their offerings include exploration of contemporary topics and applications, to communicate that mathematics today is alive and ever changing.

The phrase "strive to" in Recommendation 3 means that every course should make progress toward reaching the specified goals. Some of the goals are more accessible than others, and thus the following discussion elaborates not only the rationale for each goal but also the challenges to achieving it.

Present key ideas and concepts from a variety of perspectives. Key ideas and concepts should provide the intellectual and pragmatic themes that unify a course. These themes need to be developed from several perspectives for students to achieve the depth of understanding required for application of the material.

In recent years the undergraduate mathematics curriculum has broadened to include a greater variety of content perspectives. This trend needs to continue and deepen. The curricular emphasis in traditional mathematics instruction was on correct theoretical development of the material, with an emphasis on algorithms based on symbolic algebraic manipulation leading to exact solutions. Such a perspective on mathematics was, and still is, important. But there are other important perspectives that have been underrepresented in the traditional curriculum.

Every mathematics course—from introductory courses for liberal arts students to courses for mathematics majors—should offer opportunities for students to explore mathematical ideas from this variety of perspectives. Students differ in how they tackle problems and process information. The more varied the discussion of a topic, the better the chance that each student will find something that makes the idea understandable and memorable. This is not just good pedagogy; it enriches students' understanding of the nature of mathematics.

The theoretical perspective, based on logical accuracy in statement and verification, is a bedrock foundation for mathematics. But an overemphasis on this perspective can have negative effects on student understanding and motivation, especially in introductory courses. Developing intuitive conceptual understanding is a practical perspective that needs more exposure. Determining and achieving the proper balance is not always easy, but neither perspective can be ignored.

Developing appropriate symbolic manipulation skills will always be a goal of mathematics instruction. But in addition graphical and numerical understanding must be developed. Viewing topics from these analytical, graphical and numerical perspectives is necessary for true understanding. For example, in calculus courses students should experience geometric as well as algebraic viewpoints and approximate as well as exact solutions. In linear algebra students should learn to "see" the null space of a matrix as well as to compute a basis for it. In statistics students must recognize the geometry behind linear regression as well as how to apply it to answer meaningful questions about real data.

The traditional curriculum does not make clear the rarity of situations in which exact solutions are possible, and hence the need to balance their study with that of approximate solutions. Exact solutions are important precisely because they are rare, and this needs to be communicated. At the same time, students need to be equipped to handle situations in which exact solutions are not readily available. Evaluation of integrals is one example of a situation in which exact solutions are not common. Students need to understand the importance of the exact techniques that are available. At the same time, more emphasis is needed on proficiency with techniques of approximation.

Employ a broad range of examples and applications to illustrate and motivate the material. Examples and applications can often illustrate, explain, and justify the logical structure of a mathematical subject better than an unadorned theoretical development. Applications help students move beyond an algorithmic approach to mathematics to grapple with what concepts mean and how they are used. When students look back on a course, it is often the examples and illustrations that are most memorable. Authentic and interesting (and sometimes surprising) applications can be powerful hooks drawing a student's interest into the mathematics under study. For these reasons, every course—from the most basic to the most advanced—should include a rich assortment of examples and applications.

A growing number of textbooks now include such examples and applications. It is not necessary to be an expert in another discipline or spend substantial time providing background for students in order to incorporate them.

For instance, exposing students to numerous examples of the accumulation of a varying quantity as a generalization of a simple rule about accumulation of a constant quantity helps them understand the definite integral. Such examples include generalizing the volume of a cylinder to volumes of solids of revolution, generalizing "work equals force times distance" to work as an integral when the force varies, or generalizing simple interest calculations into an integral formulation for the present value of an income stream. Radiocarbon dating illustrates how an initial value problem (IVP) can model a real world situation, and the solution of the IVP then yields obviously useful and interesting results. It also is impressive to learn how a simple system of differential equations predicts the cyclical population swings in a predator-prey relationship, how the singular-value decomposition of a matrix facilitates the compression of information, and how modular arithmetic is used in cryptography and the transmission of encoded information.

Non-trivial applications can also lead to unanticipated mathematical questions. For example, when constructing a bio-economic model for the Maine lobster fishery based on the logistic equation, one is faced with having to model a population for which no direct population data are available: the available data consist of the tonnage of catch each year along with the number of traps in use each year. Drawing useful conclusions from these unusual data poses interesting mathematical and statistical problems.

Many examples and applications fall naturally under the rubric of mathematical modeling, and it should receive increased emphasis. Modeling requires students to translate back and forth between verbal descriptions and mathematical expressions, to make their assumptions explicit, and to decide which details are important and which should be ignored. Exposure to modeling will help students consolidate their mathematical understanding and empower them to use their understanding in other courses. Modeling also gives students practice in framing questions, trying multiple approaches, and interpreting results. Further, it provides effective topics for student writing.

According to the summary report of the Curriculum Foundations project, "The importance ascribed to mathematical modeling by *every* disciplinary group in *every* workshop was quite striking."[26] While the relative emphasis on applications will vary from one course to another, every course in the undergraduate mathematics program—from the most basic to the most advanced—should strive to include meaningful applications that genuinely advance students' ability to analyze real-life situations and construct and analyze appropriate mathematical models. It should be emphasized that modeling need not be thought of as restricted to major projects but can occur in small ways throughout a course. Often insights that instructors take for granted as "obvious" are far from obvious to their students.

Good examples don't always have to come from applications. Puzzle problems couched in everyday language also can be very effective. Solving the Tower of Hanoi puzzle for increasing numbers of disks is an excellent illustration of recursive thinking and can demonstrate the usefulness of proof by induction. Rubik's Cube is a rich source of group theory problems.

Examples can also come from mathematics itself. For instance, students may enjoy calculating pi by numerically estimating a definite integral or by finding partial sums of a series. After computing characteristic polynomials, students can use the power method to converge on the largest eigenvalue of a big matrix. And as shown in George Pólya's classic video "Let us teach guessing," students' creative imagination may be engaged in trying to determine the number of spaces created by five planes randomly positioned in space.

However, more substantial applications appropriate for instructional use are also needed, and they are neither trivial to find nor easy to frame effectively for instruction. Indeed, substantial applications—though valuable—can be quite difficult for students because they must not only learn the mathematics, but also learn the relevant areas of the discipline from which the application is drawn.

Good applications and examples should be disseminated to others in the mathematical community. Published texts and papers, such as those produced by the MAA and COMAP, provide part of the solution, and dissemination via Internet sites, such as the one for the Journal of Online Mathematics (JOMA), is of growing importance.

Published texts and papers provide part of the solution, but dissemination via Internet sites is of growing importance. The mathematical community as a whole should put more effort and resources into development

---

[26] "A Collective Vision: Voices of the Partner Disciples," *The Curriculum Foundations Project: Voices of the Partner Disciplines*, p. 4.

and dissemination of applications. And individual faculty should contribute when they can and take advantage of what is available when they cannot.

Make connections to other subjects and apply the course material to these subjects. Students often are totally unaware of important connections between separate mathematical subjects and between mathematics and other disciplines. This is a serious deficiency for both the understanding of the mathematics and for the appreciation of its importance and use.

Also, students are surprisingly reluctant or unable to apply the knowledge they obtain in mathematics courses to other disciplines. This "transfer problem" is a serious weakness in current mathematics instruction. It is not reasonable to expect students to become sophisticated modelers and applied mathematicians with a few courses nor to expect that there are guaranteed solutions to the transfer problem. However, there are strategies that are commonly understood to be helpful.

For instance, one reason students encounter difficulty in applying mathematics to problems in other disciplines is that they have trouble identifying appropriate mathematical procedures when problems are expressed with different symbols than those used in the mathematics classroom. To counter this problem, textbooks and instructors can go beyond conventional x, y notation to use a larger collection of symbols for both constants and variables. It is tempting to stay with x's and y's because extra class time is often needed to help students become accustomed to alternate symbolism. But the very fact that this is the case underscores the importance of incorporating the activity in courses.

Students' failure to recognize connections with other disciplines extends to mathematics itself: undergraduates (including mathematics majors) often see mathematics as a collection of separate, isolated subdisciplines with little interplay between the separate areas. This is a flawed and limiting picture of the discipline. The unity of mathematics must be conveyed via clearly presented connections between mathematical fields and the symbiotic interplay of theory and application.

Even upper-level courses that are traditionally limited to one corner of the mathematical landscape—abstract algebra, for example—should offer more than a narrow perspective. For instance, through symmetry groups and their subgroups, group theory is intimately tied to geometry, and applications to coding theory are elementary enough to be discussed in an introductory course. Moreover, the structures included in an abstract algebra course can be used to help students understand the basis for the mathematics they studied in elementary and secondary school. (Also see C.3 and the corresponding Illustrative Resources.)

Mathematics faculty need to be consciously aware of the need to build connections and make this a higher priority in their instruction.

Introduce contemporary topics from the mathematical sciences and their applications. Mathematics is one of the few disciplines in which many undergraduates complete degrees with little idea of what has happened in the field within the past hundred years. Consequently, although students are well aware that research is ongoing in the natural and physical sciences, they often are totally unaware that research is done in mathematics. This is a highly inaccurate and misleading perception that can decrease student interest in the field. Mathematics must be presented as a discipline with intellectual challenges and open questions, not as a subject whose vitality ended long ago. This requires presentation of contemporary topics and at least glimpses of current fields of research.

Although much modern mathematics requires extensive preparation to be fully appreciated, many contemporary topics can be included in the undergraduate mathematics curriculum, either as courses, units within courses, student projects, or simply as examples used to enliven interest. The power and ubiquity of mathematics in the modern world is an important "hook" to engage student interest. Even students taking a single course can learn that mathematics underlies technology, like CD players and secure electronic communication, that they use every day. Mathematical sciences departments have a responsibility to see that all students, definitely including mathematics majors, have opportunities to learn about contemporary topics in pure and applied mathematics.

Here is a sampling of topics that can be included in various ways and in courses at varying levels:

- Dynamical systems, including chaos and fractals. These topics can be treated in introductory, intermediate and advanced courses.

- Resampling methods. Computers have radically changed the way statisticians approach data. Statistics courses should convey to students the excitement of newly developing techniques.

- Wavelets. The common wisdom is that this topic cannot be properly presented until students have mastered Fourier analysis. However, junior/senior level undergraduate courses do exist which successfully address this material.

- Error correcting codes. Besides being excellent applications of linear and abstract algebra, error-correcting codes have important uses in new technologies such as CD-ROMs and cell phones, as well as for analyzing data-transmission from outer space.

- Complexity theory. The "P vs. NP" problem (answer 'easy to find' vs. 'easy to check') is well within the range of undergraduates' understanding. Pointing out that the Clay Mathematics Institute is offering a million dollar prize for the solution helps to pique students' interest.

Other examples include number theory and cryptography, computational algebraic geometry, computer graphics, Monte Carlo methods, game theory, knot theory, quantum computing, and algebraic combinatorics.

Thanks in some cases to judicious use of technology, new topics have become appropriate for undergraduates and existing undergraduate courses are ripe for rethinking. The examples mentioned above are only a few. Textbooks and mini-textbooks that support inclusion of these topics are just starting to appear. More are needed.

4. Promote interdisciplinary cooperation

Mathematical sciences departments should encourage and support faculty collaboration with colleagues from other departments to modify and develop mathematics courses, create joint or cooperative majors, devise undergraduate research projects, and possibly team teach courses or units within courses.

Towards Excellence urges departments to "Build strong relationships on campus. Faculty should make building strong relations with other departments and the campus administration a conscious department goal."

The Curriculum Foundations workshops demonstrated a substantial willingness of faculty in other fields to work with mathematicians to develop programs and courses and to team-teach courses or units within courses. Such interdisciplinary programs and courses are labor intensive and often costly but they are needed because of the rapidly changing nature of the disciplines that use mathematics.

Interdisciplinary programs energize both the students and faculty who participate. They entice students from other disciplines to learn more mathematics and to learn it in a context that is important to them. They open new ways of looking at mathematics for mathematics majors, including possible career options or avenues for further study. And they provide an opportunity to communicate some of the cutting-edge work of mathematics. Interdisciplinary collaborations also enrich the faculty who participate, affording them opportunities to learn powerful applications and ideas from other disciplines.

The Curriculum Foundations workshops also demonstrated that many faculty in partner disciplines are eager to provide input to help mathematicians revise existing courses to support current disciplinary needs more effectively. Communication about such courses has the added benefit of increasing awareness of their content, which improves the ability of faculty in other disciplines to build upon the mathematical backgrounds of the students in their own courses.

Joint efforts can cement relations between a mathematical sciences department and other disciplines. Lines of communication are kept open so that the mathematics department is sensitive to the needs and concerns of these partner disciplines. This has the paradoxical effect of reducing pressures on the department to conform to the wishes of others by making it easy to spot the small adjustments that can be most beneficial to them.

Effective dialogue with colleagues in other disciplines is not always easily achieved. Especially in a large institution with numerous departments requiring mathematics for their students, a mathematical sciences department may hear conflicting requests. Constrained resources limit the capacity of departments to tailor separate courses for specific partners. Priorities must be determined and compromises are necessary. But small cooperative successes can breed larger ones, and the benefits of genuine cross-disciplinary cooperation far outweigh the potential difficulties.

In order to promote the creation of interdisciplinary programs, the development of new courses, and the appropriate revision of existing ones, mathematical sciences departments should identify and encourage faculty who are particularly interested in interdisciplinary cooperation and are willing to invest the effort required to bring it about. Encouragement should include rewards such as release from other obligations, additional summer salary or professional development funds. Many institutions are suffering severe budget cuts that make funding initiatives difficult, but steady advocacy for these rewards moves them higher in the priority list when resources become more available. Moreover, even in times of budgetary constraint, departments should ensure that such activities are viewed favorably when tenure and promotion decisions are made.

When a department reaches out to and interacts with other departments, it sets an example that college and university administrators will recognize and encourage. Collaborations are more than a service to students and colleagues. They serve to enhance the visibility of a mathematical sciences department and ensure its importance in achieving the mission of the entire institution.

5. Use computer technology to support problem solving and to promote understanding

At every level of the curriculum, some courses should incorporate activities that will help all students progress in learning to use technology

- *Appropriately and effectively as a tool for solving problems;*
- *As an aid to understanding mathematical ideas.*

Mathematical sciences departments should look for opportunities to make effective use of technology—desktop and hand-held computers—at every level of the curriculum. Every instructor teaching every course should consider whether and what uses of technology are appropriate to the material and to the students' needs.

The use of technology can help students develop mathematical skills and understanding.[27] However, the use of technology must be focused on students' needs rather than on the capabilities of the technology itself. Instructors must first decide what mathematics is to be learned and how students are to learn it. The answers to these questions will determine whether and how students should use technology.

---

[27] Adapted from the 2001 reports on technology from the MAA Committee on the Teaching of Undergraduate Mathematics and the Committee on Computers in Mathematics Education and also the report of the statistics workshop of the Curriculum Foundations project.

The answers to these questions will also be affected by the technology available for student use. For example, in a beginning calculus course, the existence of graphing utilities may affect the amount of time spent on graphing by hand as well as opening the door to different approaches to graphing families of functions. Similarly, the existence of computer algebra systems can affect the treatment of techniques of integration, and the availability of a differential equation solver can permit the early introduction of modeling with systems of differential equations. As with all teaching strategies, the effectiveness of the use of technology should be evaluated throughout its implementation, and modifications should be made until the desired goals are attained. The preferences of faculty from partner disciplines should also be taken into account. For instance, participants from many disciplines at the Curriculum Foundations workshops expressed a desire for the use of spreadsheet software, when appropriate, in mathematics courses.

There should be a variety of introductory courses that make some use of technology. For example, technology can enhance student learning in introductory courses in modeling, whether with calculus or not. Most calculus students, especially those who may take only one semester, profit from the use of a graphing utility and a tool for numerical integration. In 1992 the Joint Committee on Undergraduate Statistics of the MAA and the American Statistical Association (ASA) recommended use of statistical packages for introductory statistics courses, as well as more advanced ones. Introductory (and advanced) discrete mathematics courses can take advantage of software for the manipulation of discrete graphs, theorem provers, and functional and logic-based languages.

Technology use should also be present in a variety of intermediate courses. Graphics packages enhance multivariable calculus. Most modern texts on differential equations make use of a differential equation solver. Some intermediate "bridge" courses include computer experimentation to motivate conjecture and proof. Linear algebra courses can use technology for matrix manipulation or for visualizing the effects of linear transformations in two or three dimensions.

Software exists that enhances topics in a variety of advanced courses including geometry, probability, complex analysis, algebraic geometry, and group theory. Illustrative Resources includes descriptions of effective use of technology in courses at a variety of levels and in varying institutions.

Beyond these instructional uses of computer technology, there are others. For instance, Thomas Banchoff points to the power of emerging "super texts" that use hypertext to make texts (and courses) more flexible, permitting both instructors and students to pick and choose. Wade Ellis describes the potential of tutorial software to offer remedial help to students. Gilbert Strang, who has been involved in experiments with videotaped lectures and an on-line "encyclopedia" to offer "just-in-time" help in basic mathematics courses, elaborates on the potential benefits of such strategies.[28]

In order for technology to be useful in mathematics instruction, students should be able to focus on the mathematics rather than on how to use the technology. Introductory tutorials, training sessions, and on-line help libraries can assist students to overcome difficulties with the technology. Using the same software in several different courses also can shorten the total technology learning curve. In addition, the *MAA Guidelines for Programs* specify that when a department decides to use technology in a course or program, it has a responsibility to offer appropriate training for faculty in that technology and its effective use in instruction. If resources (equipment, personnel, training) are not adequate, departments should press for their improvement.

Using technology has costs as well as benefits. Besides the obvious hardware and software costs, there also are substantial human costs, not just for support personnel (if a department is fortunate enough to have

---

[28] See "Some Predictions for the Next Decade" by T. Banchoff, "Mathematics and the Mathematical Sciences in 2010: What Should Graduates Know?" by W. Ellis, and "Teaching and Learning on the Internet," by G. Strang in *CUPM Discussion Papers about Mathematics and the Mathematical Sciences in 2010: What Should Students Know?*, MAA Reports, 2001.

them) but also for faculty who must learn to make effective use of the technology, spend time developing suitable course materials, and oversee the facilities available to students. Learning to use technology means time for students as well, time that might otherwise be spent on mathematical content. And when misused, technology can become a crutch, used for tasks that students should be able to do by hand and providing only an illusion of accomplishment.

The reality of such problems with technology should not cause mathematics departments to avoid its use. Rather, they indicate the care and effort needed for effective implementation. The potential benefits of technology for student learning are worth this care and effort.

Use technology appropriately and effectively as a tool for solving problems. Technology can help strengthen students' problem-solving skills by encouraging them to utilize multiple problem-solving strategies (graphical, numerical, algebraic); it is especially valuable for visualization. Data sets stored in local computer files or available on the Internet provide the information that makes it practical to address problems rooted in real data. This allows students to apply their mathematical knowledge to real situations, motivating them to learn and understand mathematics in the context of real-world problems. Statistical packages and spreadsheet programs free students from tedious or non-illuminating computations, perform extensive computations that are unrealistic to do by hand, produce results for students to interpret, and depict results visually as graphs, histograms, and diagrams. Technology makes large linear systems tractable in linear algebra and complicated dynamical systems accessible in introductory differential equations courses.

Students should learn to choose the appropriate means for the problem being solved. Some problems should be solved by hand. Others should be solved using technology such as calculators and spreadsheets. Still others require more substantial computer power. Students should learn to distinguish between thinking, which is a human activity, and computing, which machines do well.

Students especially need to learn how to be intelligent consumers of the answers technology provides. They should know how to make order of magnitude estimates for numerical calculation and to do trial calculations for small cases where answers can be checked by hand. They should use qualitative analysis to test quantitative results for reasonableness. For example, if approximating the integral of a positive-valued function, students should recognize that a negative answer cannot be correct.

Use technology as an aid to understanding mathematical ideas. Students should learn to visualize geometric objects, to relate graphical objects to their analytic definitions, and to see the graphical effects of varying parameters. Technology allows students easy access to the graphs of planar curves, space curves and surfaces, and more specialized software gives access to other geometric objects. Visualization also helps students understand concepts such as approximation of integrals by Riemann sums or functions by Taylor polynomials. The ability of technology to handle even symbolic manipulations allows students to focus their attention on understanding concepts. Use of technology, whether a high-level programming language or a spreadsheet, can also help students learn to think algorithmically by giving them experience working with algorithms.

Technology—especially dynamic tools—can promote students' exploration of and experimentation with mathematical ideas. For example, students can be encouraged to ask "what if?" questions, to posit conjectures, to verify or refute them, and to use technology to investigate, revise, and refine their predictions. Specific examples include studying the effects of manipulating parameters on classes of functions and fitting functional models to data. When students see that some patterns persist and others eventually fail, they begin to understand the meaning of mathematical truth, and they can learn to value proof as a source of both justification and explanation.

6. Provide faculty support for curricular and instructional improvement

Mathematical sciences departments and institutional administrators should encourage, support and reward faculty efforts to improve the efficacy of teaching and strengthen curricula.

CUPM Guide 2004 lays out a vision of excellence in the undergraduate mathematics program. The *MAA Guidelines for Programs* specify the mathematical sciences department's responsibility to promote excellence in teaching and to create programs to achieve and sustain it. Excellence can be accomplished only with sustained faculty effort, and this effort is most likely to occur in an environment where it is encouraged, supported, and rewarded.

Creating such an environment must begin with the department administrators, with deans and provosts who establish expectations for the department, and with those members of the department who enjoy prestige and influence. "Departments should ensure that senior faculty assume a leadership role in the undergraduate program by participating fully in teaching, curriculum development, and student advising."[29] These leaders should find ways to show faculty that teaching is respected and excellence in education is a departmental and institutional goal. A simple way of conveying this message is to publicize significant achievements in curriculum development and program improvement along with those in faculty research.

Department leaders should direct faculty to information about effective innovations, how they work, and practical means of implementing them. With college administrators, they should provide the necessary budgetary support to bring in speakers or workshop leaders and to make continuing professional development available to all members of the department. Adjunct and part-time faculty and graduate students are important contributors to the teaching mission of many departments. As stated in the *MAA Guidelines for Programs*, "Departments that employ part-time instructors should provide them with all of the resources necessary for teaching that are provided to full-time instructors." In particular, means should be found to provide them with mentoring, training, and professional development opportunities.

The departmental leadership should actively seek to identify and encourage those faculty who are or could be positioned to make a substantial improvement in some aspect of the educational mission of the department. This is in line with the recognition in *Towards Excellence* of "the importance of identifying the right person to lead a department initiative and giving that person the support needed to create a successful program."[30]

However, finding someone to lead an initiative is not enough. No program will long survive if it represents the work of a single individual. For long-term sustainability, initiatives must be team efforts, with faculty in supporting roles who can be prepared to expand or take over the leadership of the program. At institutions that require research for promotion and tenure, untenured faculty should not be expected to take on roles that would seriously hamper their scholarly development. However, means should be found to involve all faculty in improving the curriculum and its instruction. This is especially true for younger faculty who often bring enthusiasm and openness to innovative ideas.

There are many ways that support can be made tangible. These include release time for key faculty, clerical support, and funds for speakers, programs, or special materials. As noted earlier, advocacy for this tangible support is warranted even in times of fiscal constraint. For many effective innovations (especially those involving changes of course content or pedagogy), support includes the promotion of broad departmental engagement. Effective initiatives must be seen to be appreciated and recognized by the leadership of the department as important contributions to the mission of the department.

---

[29] *MAA Guidelines for Programs and Departments in Undergraduate Mathematical Sciences,* MAA, 2001; `www.maa.org/guidelines/guidelines.html`. See 2g.

[30] *Towards Excellence*, page 33.

The salary structure of educational institutions must respect the full range of contributions to the department's missions. Faculty should see concrete expressions of support for their educational efforts in the form of appropriate salary increases, as well as benefits such as travel allowances, funds for course development, and support for co-curricular activities. *CUPM Guide 2004* supports the recommendation of the AMS Task Force on Excellence:

"There should be clear standards of excellence for those whose greatest achievements are in teaching or other educational activities, and faculty who meet those standards should share in faculty rewards, both financially and through promotion in rank."[31]

---

[31] *Towards Excellence*, page 35.

Part II. Additional Recommendations Concerning Specific Student Audiences

A. Students taking general education or introductory collegiate courses in the mathematical sciences

General education and introductory courses enroll almost twice as many students as all other mathematics courses combined.[32] They are especially challenging to teach because they serve students with varying preparation and abilities who often come to the courses with a history of negative experiences with mathematics. Perhaps most critical is the fact that these courses affect life-long perceptions of and attitudes toward mathematics for many students—and hence many future workers and citizens. For all these reasons these courses should be viewed as an important part of the instructional program in the mathematical sciences.

This section concerns the student audience for these entry-level courses that carry college credit. A large percentage of these students are enrolled in college algebra. Traditional college algebra courses, with a primary emphasis on developing skills in algebraic computation, have a long history at many institutions.

Students enrolled in college algebra courses generally fall into three categories:

1. Students taking mathematics to satisfy a requirement but not specifically required to take a course called college algebra;

2. Students majoring in areas or studying within states or university systems that specifically require a course called college algebra;

3. Students intending to take courses such as statistics, calculus, discrete mathematics, or mathematics for prospective elementary or middle school teachers and who need additional preparation for these courses.

Unfortunately, there is often a serious mismatch between the original rationale for a college algebra requirement and the actual needs of the students who take the course. A critically important task for mathematical sciences departments at institutions with college algebra requirements is to clarify the rationale for the requirements, determine the needs of the students who take college algebra, and ensure that the department's courses are aligned with these findings (see Recommendation A.2).

---

[32] According to the CBMS study in the Fall of 2000, a total of 1,979,000 students were enrolled in courses it classified as "remedial" or "introductory" with course titles such as elementary algebra, college algebra, pre-calculus, algebra and trigonometry, finite mathematics, contemporary mathematics, quantitative reasoning. The number of students enrolled in these courses is much greater than the 676,000 enrolled in calculus I, II or III, the 264,000 enrolled in elementary statistics, or the 287,000 enrolled in all other undergraduate courses in mathematics or statistics. At some institutions, calculus courses satisfy general education requirements. Although calculus courses can and should meet the goals of Recommendation A.1, such courses are not the focus of this section.

Because many students taking introductory mathematics decide not to continue to higher level courses, general education and introductory courses often serve as students' last exposure to college mathematics. It is important, therefore, that these courses be designed to serve the future mathematical needs of such students as well as to provide a basis for further study for students who do continue in mathematics. All students, those for whom the course is terminal and those for whom it serves as a springboard, need to learn to think effectively, quantitatively and logically. Carefully-conceived courses—described variously as quantitative literacy, liberal arts mathematics, finite mathematics, college algebra with modeling, and introductory statistics—have the potential to provide all the students who take them with the mathematical experiences called for in this section.

A common feature of many effective courses and programs that have been developed for these students is the leadership provided by key faculty members. It requires committed and talented faculty to understand the needs of these students and the opportunities inherent in these courses. Continuing leadership is needed and special training must be provided for instructors—including graduate assistants and part-time faculty—to offer courses that will meet the needs of these students.

A.1. Offer suitable courses

All students meeting general education or introductory requirements in the mathematical sciences should be enrolled in courses designed to

- *Engage students in a meaningful and positive intellectual experience;*
- *Increase quantitative and logical reasoning abilities needed for informed citizenship and in the workplace;*
- *Strengthen quantitative and mathematical abilities that will be useful to students in other disciplines;*
- *Improve every student's ability to communicate quantitative ideas orally and in writing;*
- *Encourage students to take at least one additional course in the mathematical sciences.*

Different institutions will develop and offer different sets of courses designed to fit the characteristics of their student body. In recent years a large number of engaging and challenging courses have been developed and successfully offered to meet the needs of these students. However, regardless of the curriculum chosen, all general education and introductory courses should strive to meet the goals outlined in this recommendation. Each of these course goals will now be described in more detail.

Engage students in a meaningful intellectual experience. Students must learn with understanding, focusing on relatively few concepts but treating them in depth. Treating ideas in depth includes presenting each concept from multiple points of view and in progressively more sophisticated contexts. For example, students are likely to improve their understanding more by writing analyses of a single situation that combines two or three mathematical ideas than by solving half a dozen problems using each idea separately. In addition, students should encounter some meaningful ideas of mathematics. College students study the best paintings, the most glorious music, the most influential philosophy, and the greatest literature of all time. Mathematics departments can compete on that elevated playing field by offering and making accessible to all students intriguing and powerful mathematical ideas. Mathematical modeling, data analysis, quantification of the uncertain and the unknown, analysis of the infinite, cryptography, fractals—these topics and many more can compete well with any other subject for depth and fascination.[33] Indeed, these courses

---

[33] For example, if students are using geometric sequences and their sums to model a real world situation, it is natural to help them encounter the notion of infinity through an examination of the area and perimeter of the stages of a Koch snowflake.

should be developed and offered with the philosophy that the mathematical component of every student's education will contain some of the most profound and useful ideas that the student learns in college.

Increase students' quantitative and logical reasoning abilities. Departments should encourage and support their institutions in establishing a quantitative literacy program for all students, with the primary goal of developing the intellectual skills needed to deal with quantitative information as a citizen and in the workplace.[34] This program should ensure that all introductory and general education mathematics courses make a significant contribution toward increasing students' quantitative reasoning abilities. Courses in the program should be coordinated with courses in other disciplines in order to reinforce skills and place quantitative reasoning in a broader context. Clearly, no one course can enable an individual to become quantitatively literate, but introductory and general education mathematics courses can help students see how quantitative methods can be used to help answer questions that they find meaningful. Students in these courses should also have the opportunity to use a variety of mathematical strategies—seeking the essential, breaking difficult questions into component parts, looking at questions from various points of view, looking for patterns—in diverse settings. Informing the pre-college community of the institution's commitment to quantitative literacy can help work toward this goal.

Strengthen mathematical abilities that students will need in other disciplines. As reported in "A Collective Vision: Voices of the Partner Disciplines,"[35] faculty representing other disciplines emphasized the importance of mathematical modeling. Students should be able to create, analyze, and interpret basic mathematical models from informal problem statements; to argue that the models constructed are reasonable; and to use the models to provide insight into the original problem. Many of the disciplinary reports cite the importance of conceptual understanding and data analysis. A study of these reports and the textbooks and curricula of courses in other disciplines shows that the algorithmic skills that are the focus of computational college algebra courses are much less important than understanding the underlying concepts. Disciplines outside mathematics rarely ask their students to find the equation of the line that passes through two given points. But social scientists will expect students to recognize a linear pattern in a set of data, interpret the parameters of the best-fitting line, and use the equation of the line to answer questions in context. Mathematical sciences departments need to know which concepts will be needed in subsequent courses. They need to know how they will be needed. And they need to adjust their general education and introductory courses accordingly. Faculty also need to incorporate strategies to help students transfer skills learned in mathematics courses to other subjects. For instance, students should be required to attach units to answers when appropriate and to practice translating algebraic expressions into verbal ones.

Improve students' ability to communicate quantitative ideas. "A Collective Vision: Voices of the Partner Disciplines" reports that nearly every discipline promotes the importance of having students communicate mathematical and quantitative ideas—both orally and in writing. Communication skills are related to logical reasoning: if you can't explain it, you don't understand it. It is imperative that communication be emphasized in all mathematics courses. In general education and introductory courses, communication experiences should focus heavily on increasing students' quantitative reasoning abilities and making connections with the mathematics encountered in other disciplines.

Encourage students to take other courses in the mathematical sciences. On the one hand, general education courses provide the last formal mathematics experience for most students and so must stand on their

---

[34] A useful description of quantitative literacy is provided in *Mathematics and Democracy: The Case for Quantitative Literacy*, L. Steen, ed., National Council on Education and the Disciplines, 2001.

[35] The introductory article in *The Curriculum Foundations Project: Voices of the Partner Disciplines.*

own intrinsic merits. On the other hand, they should be designed to serve as gateways and enticements for other mathematics courses. No course should ever be constructed as a final mathematics course. Natural subsequent courses include other general education courses and statistics and can even include discrete mathematics or calculus. Depending upon the nature of a subsequent course, modest changes to the general education course might make it more useful as a preparation.

Each department should consider its offerings to students taking general education or introductory collegiate mathematics courses in terms of the criteria in Recommendation A.1 and its own institutional situation. It is understood that each department has unique opportunities and boundary conditions within which it must operate. These include the mathematical preparation of students, institutional admission standards, institutional transfer guidelines and articulation agreements, mathematical needs of other disciplines with large numbers of majors, institutional general education requirements, level of commitment of the institution to high quality general education, average class size, mandated state curriculum requirements and exams, interests and areas of specialization of faculty, and the extent to which instruction is provided by graduate teaching assistants and/or part-time faculty. However, every mathematics department should strive to implement the spirit of this recommendation in ways appropriate to its individual circumstances.

A.2. Examine the effectiveness of college algebra

Mathematical sciences departments at institutions with a college algebra requirement should

- *Clarify the rationale for the requirement and consult with colleagues in disciplines requiring college algebra to determine whether this course—as currently taught— meets the needs of their students;*
- *Determine the aspirations and subsequent course registration patterns of students who take college algebra;*
- *Ensure that the course the department offers to satisfy this requirement is aligned with these findings and meets the criteria described in A.1.*

Clarify the rationale. College algebra courses are specified for students in a variety of professional fields and by a number of states. To determine the actual needs of these students, mathematics faculty should consult with colleagues in disciplines requiring college algebra to determine whether this course—as currently taught—meets the needs of their students. In such discussions it is helpful to provide several textbooks representing both traditional and nontraditional approaches available because faculty in other disciplines may be unaware of the variety of materials that can be used to teach the kind of course described in A.1. Recommendations of representatives of the partner disciplines reported in *The Curriculum Foundations Project: Voices of the Partner Disciplines* indicate that a traditional, computational college algebra course does not address the needs of students in these disciplines. They recommend a course meeting the criteria in A.1. Many such courses and programs already exist. They are offered at a wide array of institutions and under a variety of names, such as college algebra with modeling, introduction to contemporary mathematics, quantitative reasoning, liberal arts mathematics, contemporary college algebra, and pre-calculus and modeling. Short descriptions of sample courses are provided in Illustrative Resources, along with a list of possible textbooks.

Determine student needs. This is a restatement of Recommendation 1 for the special case of students taking college algebra. What are the mathematical backgrounds of these students? What are their intended majors? Do they take courses for which college algebra is a prerequisite? Do they actually *use* what they learned in college algebra in subsequent courses? Are they successful in those courses? If, for example, the addition of rational expressions is never used, why is it appropriate for these students? Is it an idea of intrinsic interest and value?

Align the course with these findings. Students should indeed be equipped with the specific mathematical skills they need for their academic programs. But they should also be equipped with mathematical knowledge that will remain with them and be useful in their careers and in their lives as contributing citizens. The criteria in A.1 are an appropriate basis for the design of this course.

A.3. Ensure the effectiveness of introductory courses

General education and introductory courses in the mathematical sciences should be designed to provide appropriate preparation for students taking subsequent courses, such as calculus, statistics, discrete mathematics, or mathematics for elementary school teachers. In particular, departments should

- ***Determine whether students who enroll in subsequent mathematics courses succeed in those courses and, if success rates are low, revise introductory courses to articulate more effectively with subsequent courses;***
- ***Use advising, placement tests, or changes in general education requirements to encourage students to choose a course appropriate to their academic and career goals.***

Determine whether students succeed in subsequent courses. For example, students who will subsequently study statistics should be prepared to undertake a data-driven course that emphasizes statistical thinking and contributes to their quantitative literacy. According to the recommendation from the statistics report in *The Curriculum Foundations Project: Voices of the Partner Disciplines,* students who are taking mathematics to prepare for the study of statistics need to "develop skills and habits of mind for problem solving and for generalization" (p. 125). The workshop participants recommended that mathematics courses for students preparing to study statistics should "emphasize multiple approaches for problem solving,... insist that students communicate in writing,... [and use] real, engaging applications through which students can learn to draw connections between the language of mathematics and the context of the application" (p. 125).

A description of a statistics course that fully meets Recommendation A.1 is provided in a position paper presented in August of 2000 at the American Statistical Association's Undergraduate Statistics Education Initiative.[36] While such a course does not necessarily have any college-level prerequisites, many institutions successfully offer a two-course sequence that includes a mathematics course and a statistics course both of which meet the goals of Recommendation A.1.

For another example, students who are preparing to study calculus need to develop conceptual understanding as well as computational skills. Appropriately designed pre-calculus courses can enable students to be successful in calculus. Often, creation of an effective pre-calculus course requires learning about different curricular and pedagogical approaches and experimenting with how the most promising ones might be adapted for local implementation. Students who are preparing to study the entry-level discrete mathematics course now recommended for computer science students by the computer science societies need essentially the same mathematical background as students preparing to study calculus, except that trigonometry is not normally needed for discrete mathematics.

The discussion here concentrates on preparation for subsequent courses, but it should be remembered that students' plans change, and an introductory course could end up being a student's last formal exposure to mathematics. No course should have value *only* as a preparation for a subsequent course. It should have intrinsic value on its own as well as offering preparation for further study.

---

[36] Garfield, Joan, Bob Hogg,, Candace Schau, and Dex Whittinghill, (2002), "First Courses in Statistical Science: The Status of Educational Reform Efforts," *Journal of Statistics Education,* Volume 10, Number 2, www.amstat.org/publications/jse/v10n2/garfield.html.

Help students choose appropriate courses. For example, pre-calculus is frequently chosen inappropriately. Ordinarily, students not intending to study calculus should be discouraged from registering for pre-calculus and encouraged to choose courses more appropriate for their future mathematical needs. Carrying out this recommendation requires liaison work with other departments to improve placement and to clarify the true needs of their programs. Similarly, appropriate placement in other introductory courses requires conferring with departments whose students take those courses, educating academic advisors, and perhaps making use of placement tests. In institutions where students' academic plans are very fluid in the first two years, it is especially important that students take introductory mathematics courses that will flexibly equip them for future study of mathematics should their plans change.

B. Students majoring in partner disciplines

Partner disciplines vary by institution but usually include the physical sciences, the life sciences, computer science, engineering, economics, business, education, and often several social sciences.[37] It is especially important that departments offer appropriate programs of study for students preparing to teach elementary and middle school mathematics. Recommendation B.4 is specifically for these prospective teachers.

Many of the courses taken by students majoring in the partner disciplines are also taken by students who may choose to major in the mathematical sciences. (Sometimes they are the same students.) Obviously, the needs of these prospective mathematical sciences majors also warrant careful consideration.

The recommendations in this section are heavily influenced by the findings of the Curriculum Foundations project (see Appendix 2). The resulting reports, published in *The Curriculum Foundations Project*: *Voices of the Partner Disciplines*, can assist mathematical sciences faculty in discussions with faculty from other departments, serving both as a guide and a resource for collaboration.

B.1. Promote interdisciplinary collaboration
Mathematical sciences departments should establish ongoing collaborations with disciplines that require their majors to take one or more courses in the mathematical sciences. These collaborations should be used to
- ***Ensure that mathematical sciences faculty cooperate actively with faculty in partner disciplines to strengthen courses that primarily serve the needs of those disciplines;***
- ***Determine which computational techniques should be included in courses for students in partner disciplines;***
- ***Develop new courses to support student understanding of recent developments in partner disciplines;***
- ***Determine appropriate uses of technology in courses for students in partner disciplines;***
- ***Develop applications for mathematics classes and undergraduate research projects to help students transfer to their own disciplines the skills learned in mathematics courses;***
- ***Explore the creation of joint and interdisciplinary majors.***

The explosive growth and development of scientific disciplines over the past half century have resulted in unprecedented pressures on curriculum. Partner discipline faculty want to provide students with the necessary background and tools to understand current developments, but degree programs ordinarily limit

---

[37]See Appendix 2 for a list of the disciplines represented at the Curriculum Foundations workshops.

offerings to the same number of courses required fifty years ago. As a result, more topics and more sophisticated material are packed into courses of partner disciplines, magnifying the reluctance to increase allied field requirements in mathematics.

Yet many recent developments in partner disciplines are based on mathematics that is not presently included in the required courses. For example, the report of the National Research Council, *BIO 2010: Transforming Undergraduate Education for Future Research Biologists,* notes that most biology majors study calculus, and some may take a statistics course. However, the report goes on to say that these students would benefit greatly from also studying "discrete mathematics, linear algebra, probability, and modeling." The report notes, "While calculus remains an important topic for future biologists, the committee does not believe biology students should study calculus to the exclusion of other types of mathematics."[38]

When faculty teach a course in one of the partner disciplines that could use mathematics beyond that required for their majors, what do they do? Typically, either they incorporate their own brief introduction to the mathematical material or they decrease the mathematical content of their course. Both accommodations deprive students of the opportunity to deepen their scientific knowledge. Ultimately, such decisions adversely affect the health of all disciplines.

Programs for pre-service teachers are also under pressure. The increased technological sophistication of the world we live in has created a need for a more mathematically and technologically literate citizenry. Preparing future teachers to help address this need is a critical challenge for mathematics departments.

Mathematics courses, textbooks, and curricula changed dramatically during the twentieth century, often in response to the physics and engineering of the time. Further changes are needed in the twenty-first century to respond to the needs of the expanding set of disciplines for which mathematics now plays an increasingly important role. Strategies for initiating productive conversations with faculty in partner disciplines to foster such changes include:

1. visiting the courses of colleagues in partner disciplines, and discussing observations,
2. requiring a writing component within existing mathematics courses, and using this requirement as an opportunity to work with technical writing faculty,
3. inviting colleagues in partner disciplines to discuss curricular issues over lunch or at department social gatherings,
4. expanding the boundaries of faculty research to encourage collaboration with partner disciplines.

As noted in the discussion of Recommendation 4, constrained resources limit the capacity of departments to tailor separate courses for specific partners, so compromises are necessary. Nonetheless, the benefits of interdisciplinary cooperation far outweigh the potential difficulties.

At most institutions, only a portion of the mathematics faculty is interested in vigorously pursuing collaboration with faculty from other disciplines on course development and cooperative teaching arrangements. It is vital that departments support this activity with appropriate professional development (including opportunities for graduate student, part-time, and temporary teaching staff). As stated in the *MAA Guidelines for Programs and Departments in Undergraduate Mathematical Sciences,* "Participation in programs designed to assist college teachers is particularly important for members of a department who sometimes teach outside of their own mathematical sciences discipline." The *Guidelines* also say, under

---

[38] The report of the National Research Council, *BIO 2010: Transforming Undergraduate Education for Future Research Biologists*, The National Academies Press, 2002, contains an excellent description of the need for interdisciplinary education for future life science researchers. The description of the type of mathematics that these students need to study is from page 5 of the Executive Summary. The report is available at `books.nap.edu/books/0309085357/html/5.html`.

Faculty Evaluation and Rewards, "In accordance with departmental mission and priorities, some consulting and other professional activities may advance the scholarship and teaching of faculty members and the department."[39] For some faculty, interdisciplinary collaboration on course development and cooperative teaching constitutes such a professional activity, and their departments should therefore consider this activity positively in recommendations for tenure, promotion, and financial compensation. It also is important to the future vitality of the department that interest in and aptitude for such collaborations be regarded as a plus when departments consider hiring new faculty.

Strengthen existing courses that primarily serve the needs of partner disciplines. Existing mathematics courses should be carefully rethought in order to better serve the current mathematical needs of partner disciplines. As an example, several representatives of partner disciplines at the Curriculum Foundations workshops expressed a desire for greater emphasis on differential equations in the first year, while others indicated a desire for matrix algebra through eigenvalues and eigenvectors. Most representatives indicated some desire for mathematics courses that incorporate appropriate technical tools, and all wanted greater emphasis on modeling and conceptual understanding of mathematical topics.

The high priority placed on modeling and conceptual understanding is not surprising. The partner disciplines need students who can apply mathematics to questions in their fields, reformulate such questions using the appropriate mathematical tools, and use appropriate technology to carry out actual computations. Students can perform these tasks accurately and confidently only if they understand the related mathematical structure.

Representatives from chemistry gave an urgent plea for an earlier introduction of multivariable calculus because multidimensional topics underlie concepts in introductory chemistry courses. Even more colleagues expressed dissatisfaction with the level of student understanding of concepts—geometric and otherwise—in three dimensions. Specifically, they stressed the need for early introduction of vectors in two and three dimensions, geometric and graphical reasoning, linear systems, and three-dimensional visualization skills.

An important lesson from the discussions with representatives of other disciplines at the Curriculum Foundations workshops is that different disciplines use language in different ways, and it is essential to be specific to avoid misunderstanding what the issues are. Moreover, constrained resources may force compromises. But experiences at several institutions, large and small, indicate that communication opens the door to progress.

At some institutions, mathematics courses to support the programs of partner disciplines are taught within those disciplines—for instance, statistics in biology departments and discrete mathematics in computer science departments. When this is the case, mathematical sciences departments can approach these departments to signal a willingness to be involved in the further development of the courses. Such an overture may be especially welcome when the partner departments are having difficulty staffing the courses. Often the reason the courses were originally developed was because mathematicians were not interested or were not teaching the topics in a way or at a level that was appropriate for the students in the partner discipline.

Unfortunately, students who take mathematics courses outside the mathematics department are unlikely to think of taking follow-up courses within mathematics, even though such a course might enhance their understanding of the other discipline. Regardless, introductory courses housed in partner departments should be carefully examined to determine whether they can reasonably be allowed as prerequisites for higher-level mathematics courses. Mathematics faculty should attempt to publicize the mathematics courses for which courses taught in other departments can serve as preparation.

---

[39] *MAA Guidelines for Programs and Departments in Undergraduate Mathematical Sciences*, MAA, 2001; www.maa.org/ guidelines/guidelines.html. See C.2.e.f and C.8.e.

Determine appropriate computational techniques. New topics and greater numbers of conceptual and modeling problems cannot be successfully added to a fixed-length course without eliminating or reformulating existing material. Curriculum Foundations workshop participants provided insight about the kinds of topics and problems that might be de-emphasized or eliminated.

It clearly is not appropriate for mathematics departments to be cavalier in reducing the teaching of computational skills. Students must develop a certain level of computational skill in order to understand mathematics and apply it effectively. But the specific skills that are important have become less clear since the advent of technology.[40] Careful analysis by mathematics faculty, working in concert with faculty from partner disciplines, can help determine how best to modify existing courses. Joint examination of problem sets, textbook assignments, and examinations—as well as cross-disciplinary classroom observations—can lead to agreements about essential mathematical material. Based on comments made by Curriculum Foundations workshop participants, mathematics faculty may well be surprised by the views of their colleagues about what topics are inessential.

Certainly the views of colleagues in partner disciplines should not overwhelm all other considerations. Specifically, mathematics faculty can speak to what is pedagogically feasible in teaching mathematics—especially concerning the material one can expect students to absorb in a fixed amount of time. Moreover, the content of these courses must also serve the needs of the prospective mathematics majors who are enrolled. The spirit of this recommendation is to involve both mathematics and non-mathematics faculty in discussions about modifying courses, keeping open minds and flexible attitudes about various possibilities.

Develop new courses. Reports from the Curriculum Foundations workshops emphasize the importance of an elementary, non-calculus-based **statistics and data analysis** course developed through careful consultation with faculty from partner disciplines. Comments from participants suggest that such a course would be an attractive addition to their major course requirements—even if it increased the total hours required for graduation. With today's computing tools, instructors can use real data sets from a variety of applications, websites, and information from popular newspapers and magazines to motivate topics and illustrate concepts. Use of real data helps students understand the need for random sampling in survey work and the principles of designing good experiments. Students can be introduced to the difficulties of collecting meaningful data and experience the "messiness" that often accompanies data analysis, such as that involved in decisions about whether to include an outlier in an analysis and how to measure the influence of an individual observation. Analyzing real data also serves as a constant reminder of the questions that led to the original data collection.

The report from computer science indicates that entry-level, non-calculus-based **discrete mathematics** —including a serious introduction to logic and proof, sets, relations, and functions—is essential for computer science majors. In 2001, the two major computer science societies (Association for Computing Machinery (ACM) and the Computer Science Division of the Institute of Electrical and Electronics Engineers (IEEE)) released recommendations that included discrete mathematics as part of computer science "core knowledge." These societies' report, *Computing Curriculum 2001*,[41] recommended at least one discrete mathematics course in the first year, and preferably a two-course sequence. Mathematical sciences departments that currently do not offer a course satisfying this recommendation should meet with colleagues in computer science to discuss options, utilizing team-teaching when possible. A discrete mathe-

---

[40]For instance, physics workshop participants wrote: "…effective introductory physics instruction requires that students have complete confidence in their ability to understand and calculate simple derivatives and integrals. The more esoteric and complicated topics are not of use in the introductory courses. Furthermore, they are forgotten (and must be re-learned) by the time students reach upper-level courses." (*The Curriculum Foundations Project: Voices of the Partner Disciplines*, p. 116.)

[41]*Computing Curriculum 2001* is available at `www.computer.org/education/cc2001/final/`.

matics course or course sequence can be developed to serve mathematics majors, as well as computer science students, by smoothing the transition from calculus-level mathematics to more theoretical and abstract upper-division courses.

Course development in partnership with other disciplines is also important at the **intermediate and upper levels**. Especially in connection with joint and interdisciplinary majors, mathematics faculty may consider working with faculty in other areas to add courses such as mathematical biology, computational mathematics, the geometry of computer graphics, and financial mathematics.

Determine appropriate uses of technology. Participants in the Curriculum Foundations workshops understood the importance of technology, and took for granted that mathematics courses would incorporate technology to some degree. Mathematics courses should teach students how to choose the appropriate method of calculation for a given task—whether mental, paper-and-pencil, or technological—and should stress intelligent and careful interpretation of results.

Perhaps more surprising is that spreadsheets are the most utilized technology for a large number of partner disciplines. Although individual workshop reports stopped short of recommending spreadsheets as the primary technological tool in mathematics instruction, their widespread use is relevant to the technology choices made in mathematics courses that primarily serve other disciplines. A related observation was that very few workshop participants reported the use of graphing calculators in their own disciplines. Therefore, the use of calculators in mathematics courses should be for pedagogical, logistical, or budgetary reasons, not to support the use of that particular hardware in other disciplines. The bottom line: mathematics faculty need to be aware of the preferred tools of the partner disciplines in general and at their own institutions.

Develop applications and undergraduate research projects. Students often have difficulty seeing the relationships between problems in non-mathematical disciplines and material studied in mathematics courses. Interdisciplinary cooperation can help students begin to overcome this transfer problem from mathematics courses to partner discipline courses. Most faculty in partner disciplines believe that exposing students in mathematics courses to discipline-specific contexts for various mathematical topics has a positive effect on their ability to transfer knowledge between courses.

Indeed, the participants in the Curriculum Foundations workshops were so excited by the possibility of increasing the use of real models in mathematics courses that many volunteered to help develop such models. This enthusiasm suggests that if mathematicians at individual institutions invite colleagues to help develop disciplinary applications for mathematics courses, there is a good chance that they will receive a positive response. However, it is the responsibility of mathematics faculty to take the initiative in seeking such collaborative arrangements.

Interdisciplinary lectures—in both mathematics and partner discipline courses—also can serve as a means for increasing students' abilities to transfer knowledge between disciplines. An engineer or economist presenting specific applications to a mathematics class increases the sense of relevance for students. Conversely, a mathematics professor coming into an engineering or economics course when specific mathematics topics are being reintroduced reinforces earlier mathematics instruction (see Illustrative Resources for examples).

Explore creation of joint and interdisciplinary majors. Many students in fields such as biology, computer science, or economics do not think of pursuing a double major with mathematics, either because the course requirement total is too great or because they think of a mathematics major as not relevant to their career path. Joint and interdisciplinary majors are increasingly important avenues for such students to prepare themselves for advanced work in the newer, more mathematically intensive parts of their disciplines.

By working together, mathematics faculty and faculty in partner disciplines can develop programs that are specially tailored to appeal to students whose primary interests lie outside mathematics but who enjoy mathematics and can benefit from more mathematics than is required in their major. Such programs often require more courses than the combination of a major in the other discipline and a minor in mathematics, but fewer courses than a double major.

Joint and interdisciplinary majors have the double benefit of improving the mathematical knowledge of students in partner disciplines and increasing enrollment in intermediate and upper-level mathematics courses. In some cases, these programs utilize existing mathematics courses, without any alteration. In other cases, modifications may be considered, such as adding examples and applications. Mathematics faculty should be open to these kinds of modifications, recognizing that mathematics majors also benefit from learning more about the relationships between mathematical structures and real-world models. Occasionally, the introduction of joint or interdisciplinary majors may require development of entirely new mathematics courses. Such new courses are a positive sign of the vitality of the field, as well as an opportunity for professional growth (see Illustrative Resources for examples).

B.2. Develop mathematical thinking and communication

Courses that primarily serve students in partner disciplines should incorporate activities designed to advance students' progress in

- *Creating, solving, and interpreting basic mathematical models;*
- *Making sound arguments based on mathematical reasoning and/or careful analysis of data;*
- *Effectively communicating the substance and meaning of mathematical problems and solutions.*

Although mathematical thinking and communication have already been discussed (see Recommendation 2), students majoring in the partner disciplines have additional needs in this area. The partner disciplines value the precise, logical thinking that is an integral part of mathematics. Faculty who participated in the Curriculum Foundations workshops commented frequently that all disciplines look to mathematics courses to enhance students' abilities to reason logically and deductively, but that they want this ability developed in a context that increases understanding of underlying concepts. While faculty from disciplines other than computer science were wary about the use of formal proof, especially in early courses, the computer scientists specifically requested that lower-division discrete mathematics courses include an introduction to formal proof.

Different disciplines may require different levels and types of precision and logical thinking. For instance, business and economics often require qualitative analysis and a broad overview, while engineering may require more detailed and formal analysis. The natural sciences may require heuristic arguments and drawing sound conclusions from data, while computer science and software engineering require an ability to use logic to do simple proofs.

Logical, deductive reasoning and rigorous proof are at the heart of mathematics. It is important that these fundamental aspects of mathematics are embedded in the undergraduate curriculum, while at the same time not burdening the courses that serve partner disciplines with an overemphasis on formal proof. The correct balance can and should be determined through consultation with colleagues in partner disciplines.

B.3. Critically examine course prerequisites

Mathematical topics and courses should be offered with as few prerequisites as feasible so that they are accessible to students majoring in other disciplines or who have not yet chosen majors. This may require modifying existing courses or creating new ones. In particular,

- *Some courses in statistics and discrete mathematics should be offered without a calculus prerequisite;*
- *Three-dimensional topics should be included in first-year courses;*

- *Prerequisites other than calculus should be considered for intermediate and advanced non-calculus-based mathematics courses.*

As much as possible, prerequisite requirements for mathematics courses should be designed to make the courses accessible to students majoring in other disciplines and to students who have not yet chosen majors. For instance, many institutions now offer a version of calculus that incorporates review of precalculus along with calculus topics. Other institutions have found that material developed through the calculus renewal movement enables students to master the basic concepts and application of calculus, despite limitations of algebra skills that might have impeded their success in a more traditional course.

Statistics and discrete mathematics. Faculty in partner disciplines do not want a calculus prerequisite for introductory statistics. The fundamental ideas of statistics, such as the omnipresence of variability and the ability to quantify and predict it, are important subjects that can be studied without sophisticated mathematical formulations. In particular, the notion of sampling distribution—which underlies the concepts of significance testing and confidence interval—is challenging enough on its own to justify a first course in statistics. Similarly, computer scientists have requested that the introductory course in discrete mathematics be accessible to computer science students in their first year, and a number of mathematics departments now offer such a course or course sequence.

Three-dimensional topics. As noted earlier, faculty in several partner disciplines want greater emphasis on vectors in two and three dimensions, geometric and graphical reasoning, linear systems, and three-dimensional visualization skills. None of these topics require a full year of calculus to be understood, and all can be embedded in first-year courses in a variety of ways. Some institutions have structured their calculus sequence so that multivariable calculus can be reached in the second semester rather than having to wait until the third. This approach is particularly useful to students intending to major in chemistry or economics.

Advanced courses. One of the most serious consequences of overly rigid prerequisite structures is that they unnecessarily deprive mathematical sciences departments of potential students for intermediate and advanced mathematics courses. Some prerequisites are necessary, but generally not as many as are typically required. For example, if the only real requirement for a certain course is "mathematical maturity," then courses other than calculus may provide it. Many versions of discrete mathematics courses stress development of students' ability to prove and disprove mathematical statements, and some even contain brief introductions to number theory, combinatorics, or matrix theory. Such a course could be accepted as a prerequisite for linear algebra, number theory, abstract algebra, or combinatorics. Mathematics faculty should also explore allowing mathematics courses taught in other departments to serve as prerequisites for their own courses (see section B.1).

B.4. Pre-service elementary (K–4) and middle school (5–8) teachers

Mathematical sciences departments should create programs of study for pre-service elementary and middle school teachers that help students develop

- *A solid knowledge—at a level above the highest grade certified—of the following mathematical topics: number and operations, algebra and functions, geometry and measurement, data analysis and statistics and probability;*
- *Mathematical thinking and communication skills, including knowledge of a broad range of explanations and examples, good logical and quantitative reasoning skills, and facility in separating and reconnecting the component parts of concepts and methods;*
- *An understanding of and experience with the uses of mathematics in a variety of areas;*
- *The knowledge, confidence, and motivation to pursue career-long professional mathematical growth.*

The teacher preparation recommendations of this *Guide* have been informed by *The Mathematical Education of Teachers (MET),* [42] a recent CBMS report that presents detailed and carefully considered guidelines concerning the education of future teachers of mathematics. Mathematics faculty and departments are advised to study *MET* in its entirety.[43]

The need for sound mathematical preparation of prospective teachers is acute.

> The predicted huge shortage of mathematically trained teachers at all grade levels, the growing levels of state-mandated assessment and testing, and the perceived weak preparation in mathematics of entering college freshmen all point to the fact that teacher preparation must be one of the highest priorities of all institutions of higher education, and especially the state-supported ones.[44]

Mathematical sciences departments in institutions that offer programs leading to elementary and/or middle school certification face a substantial challenge, and the stakes could not be higher. These departments bear the major responsibility to provide future teachers with the solid knowledge of mathematics they require. But every mathematical sciences department also has a role to play. In a recent report, the Committee on Science and Mathematics Teacher Preparation of the National Research Council reminds higher education faculty that:

> ... all colleges and universities, including those which do not have formal teacher education programs, should become more involved with improving teacher education because the nation's teacher workforce consists of many individuals who have matriculated at all types of two- and four-year institutions of higher education. Although many of these schools do not offer formal teacher education programs, virtually every institution of higher education, through the kinds of courses it offers, the teaching it models, and the advising it provides to students, has the potential to influence whether or not its graduates will pursue careers in teaching.... Science, mathematics, and engineering departments at two- and four-year colleges and universities should assume greater responsibility for offering college-level courses that provide teachers with strong exposure to appropriate content and that model the kinds of pedagogical approaches appropriate for teaching that content.[45]

The phrase "solid knowledge of mathematics" means mathematical training that promotes real understanding of mathematical concepts at a level needed to be able to teach them effectively. This statement presents a risk and a challenge.

First, the phrase "solid knowledge of mathematics" does not mean the same thing as preparation for the further study of college-level mathematics. For example, while prospective teachers need knowledge of algebra, as detailed in the next subsection, the traditional college algebra course with a primary emphasis on developing algebraic skills does *not* meet the needs of elementary teachers as described in B.4 and *MET*.

Second, much work remains to be done in specifying exactly *what* underlying mathematics is required. Examples are needed of problems for pre-service teachers that can be used to improve students' ability to

---

[42] *The Mathematical Education of Teachers*, volume 11 of the Issues in Mathematics Education series of the Conference Board of the Mathematical Sciences, AMS and MAA, 2001, available at www.cbmsweb.org.

[43] The *MET* authors write, "This report is not aligned with a particular school mathematics curriculum, although it is consistent with the National Council of Teachers of Mathematics' *Principles and Standards for School Mathematics* as well as other recent national reports on school mathematics."

[44] "The Mathematics Major Overview," by D. Sánchez, in *CUPM Discussion Papers about Mathematics and the Mathematical Sciences in 2010: What Should Students Know?,* MAA Reports, 2001, p. 82.

[45] National Research Council. *Educating Teachers of Science, Mathematics, and Technology: New Practices for the New Millennium*. Washington, DC: National Academy Press, 2001, pp. 9–12.

reason mathematically and to evaluate children's thinking. It is necessary to know what foundational knowledge pre-service teachers need to learn in their content courses to equip them to judge the validity of their students' arguments/justifications and work. And determining what algebraic concepts and skills are essential for elementary teachers to know and be able to use is still an issue not resolved—but is one that urgently needs to be addressed.

Mathematics faculty should collaborate with mathematics educators in the school (or department) of education to maximize the effectiveness of their efforts. Two-year college faculty are important in the preparation of teachers; CBMS 2000 data (pp. 125–126) show that about half of all two-year colleges offer a special course for pre-service K–8 teachers, and enrollments rose 13% in these courses between the 1995 and 2000 CBMS reports. Given the diversity in content, number of hours, and number of courses offered at four-year and at two-year colleges, collaboration is needed to effect smooth articulation, as well as coherence, in the mathematical preparation of prospective teachers. "Greater communication and cooperation is necessary among all stakeholders in the mathematics preparation of teachers."[46]

According to *MET*, adequate development of the abilities specified in Recommendation B.4 requires at least the equivalent of 9 semester hours of appropriate mathematics courses for pre-service elementary teachers, and 21 semester hours for students who will (or may at some time) teach mathematics at the middle school level. Special mathematics courses for pre-service elementary school teachers have a long-established history, and several innovative approaches are currently being developed to improve them. In addition, many institutions are now creating courses and minors for pre-service middle school mathematics specialists because courses designed primarily for mathematics majors are not likely to reach or effectively serve these students.

Middle school mathematics (grades 5–8). Nationwide there is a shortage of teachers prepared to teach middle school mathematics. As a result school systems have been forced to assign individuals to middle school mathematics classes who originally prepared to teach at the elementary level or who prepared to teach another subject in middle school. According to the National Center for Education Statistics,[47] of those middle school teachers for whom mathematics is their *main assignment*, 53.1% do not have the equivalent of either a major *or minor* in mathematics or mathematics education.

Many of these teachers therefore do not have a sufficient understanding of what they are teaching and compensate by teaching mathematics as a set of rules or procedures not as a way of thinking (a hazard at any level of instruction). The *MET* discussion of middle grade mathematics highlights the following conclusion of L. Resnick, "Good mathematics learners expect to make sense of the rules they are taught…"[48] It is precisely this habit and desire to "make sense" that future middle school teachers need to develop and subsequently to share with their students.

While no department can meet every challenge there is a massive need for well-prepared middle school teachers. This *Guide* urges departments, particularly those in colleges that prepare substantial numbers of teachers, to develop and offer programs that will attract large numbers of potential middle school mathematics teachers. Some departments have designed minors in mathematics to prepare these students appropriately. See the Illustrative Resources for details.

Elementary mathematics (grades K–4). The mathematical preparation of elementary school teachers poses its own unique challenges. Since elementary school teachers are usually generalists, teaching all sub-

---

[46]*The Curriculum Foundations Project: Voices of the Partner Disciplines*, p. 145.

[47]National Center for Education Statistics NCES 2002-603.

[48]Resnick, L. B. (1986). The development of mathematical intuition. In M. Perlmutter (Ed.), *Perspectives on intellectual development: The Minnesota Symposia on Child Psychology* (Vol. 19, pp. 159–194). Hillsdale, NJ: Erlbaum.

jects in the elementary curriculum, many pre-service elementary teachers do not approach their future career with any particular interest in mathematics. Indeed, "... many teachers were convinced by their own schooling that mathematics is a succession of disparate facts, definitions, and computational procedures to be memorized piecemeal. As a consequence, they are ill-equipped to offer a different, more thoughtful kind of mathematics instruction to their students." (*MET*, p. 17)

There needs to be greater awareness that elementary mathematics is rich in important ideas and that its instruction requires far more than simply knowing the "math facts" and a handful of algorithms. "It is during their elementary years that young children begin to lay down those habits of reasoning upon which later achievement in mathematics will crucially depend. ....When the goal of instruction is to help children attain both computational proficiency and conceptual understanding, teaching elementary school mathematics can be intellectually challenging." (*MET*, p. 15) Mathematics programs for prospective K–4 teachers must emphasize the intellectual depth of the elementary mathematics curriculum and provide the pedagogical tools to effectively teach this critical material to elementary school children.

A solid knowledge of mathematical topics at a level encompassing much more than computational or symbolic fluency is essential for teachers of mathematics at any level. Teachers must be "able to represent concepts in multiple ways, explain why procedures work, or recognize how two ideas are related." They must also be "able to solve problems and to make connections among mathematical topics or between mathematics and other disciplines."[49]

- The study of **number and operations** provides opportunities for prospective elementary teachers "to create meaning for what many had only committed to memory but never really understood." (*MET*, p. 18) Middle school teachers need to understand the algorithms for multiplication and division of integers, rational numbers and decimals well enough to teach them "in ways that help their students remember them without resorting to thoughtless, rote techniques and that serve as a foundation for later learning." (*MET*, p. 29)

- Although formal study of **algebra and functions** is not part of the elementary curriculum, state and national standards and the National Assessment of Educational Progress (NAEP) tests require algebraic thinking from the early grades on. Prospective elementary teachers should be able to represent and justify general arithmetic claims, recognize properties of operations, appreciate that a small set of rules governs arithmetic, and be able to represent and interpret functions by graphs, formulas and tables. (*MET*, p. 20) Middle school teachers need to be able to work with algebra as a symbolic language, a problem-solving tool, and as generalized arithmetic. They need to understand and use variables and functions, especially linear, quadratic, and exponential functions. They should be able to give a rationale for common algebraic procedures. (*MET*, p. 30)

- **Geometry and measurement** have become more prominent in the elementary grades, and prospective elementary teachers need competence in visualization skill, understanding of basic shapes and their properties, and the process of measurement. Teachers in the middle grades also need to understand congruence and similarity, be able to make conjectures and prove or disprove them, understand, derive and use measurement techniques and formulas, and connect geometry to other topics. (*MET*, p. 32)

- **Data analysis and statistics and probability** are new to most prospective elementary teachers. They need to learn to pose questions that can be addressed by data, design and conduct data investigations, represent data in multiple ways, interpret results, and make judgments under conditions of uncertain-

---

[49] *The Curriculum Foundations Project: Voices of the Partner Disciplines*, p. 147.

ty. (*MET*, p.23) Middle school mathematics teachers need to understand random sampling or random assignment to treatments, explore and interpret data by observing patterns and departures from patterns, and understand what it means to draw conclusions with measures of uncertainty.

Mathematical thinking and communication skills are vital for teachers. Their needs are well served by the guidelines given in Recommendation 2, but in some respects they must go farther to acquire the abilities necessary to recognize and shape the mathematical thinking of their students. In order for prospective teachers to move beyond the attitudes and strategies they often bring from their own experiences, "they need to have classroom experiences in which they become reasoners, conjecturers, and problem solvers." (*MET*, p. 56)

The uses of mathematics should be included in the teaching of mathematics at every level. Recommendation 3 applies to courses for prospective teachers, as it does to all mathematics courses, and, suitably interpreted, it also applies to the teaching of school mathematics. Teachers need a wide repertoire of examples that illustrate the power of mathematics. Without exposure to significant applications a teacher will be effectively crippled, unable to understand accurately—and hence unable to convey—the primary motivations for many K–8 mathematics topics. It is a challenge to develop the solid knowledge of mathematical topics specified in B.4 in just three courses focused on the K–8 curriculum. However, employing applications to motivate and illustrate the material helps sustain students' interest and assists them in building that solid knowledge. Experiences with mathematical modeling afford excellent opportunities for translating between mathematical and verbal description, clarifying assumptions, and interpreting results.

Career-long professional mathematical growth is necessary to achieve and maintain excellence in the mathematics classroom. "In some countries where student achievement is high, teachers, alone and in groups, spend time refining their lessons and studying the underlying mathematics." (*MET*, p. l0) College mathematics faculty have little power to create these opportunities for professional development, but faculty can ensure that prospective teachers' college experiences in learning mathematics prepare and motivate them to take advantage of future opportunities to strengthen their understanding of mathematics.

C. Students majoring in the mathematical sciences

The recommendations in this section refer to all major programs in the mathematical sciences, including programs in mathematics, applied mathematics, and various tracks within the mathematical sciences such as operations research or statistics. Also included are programs designed for prospective mathematics teachers, whether they are "mathematics" or "mathematics education" programs, although requirements in education are not specified in this section.

Although these recommendations do not specifically address minors in the mathematical sciences, departments should be alert to opportunities to meet student needs by creating minor programs—for example, for students preparing to teach mathematics in the middle grades.

These recommendations also provide a basis for discussion with colleagues in other departments about possible joint majors with any of the physical, life, social or applied sciences.

Two premises underlie the recommendations in this section: the number of bachelor's degrees granted in the mathematical sciences—including joint majors— should be increased, and the population of potential majors has changed.

The number of bachelor's degrees granted in the mathematical sciences should be increased.

- Many other disciplines, including computer science, business, economics, the life sciences, medicine, the physical sciences, and engineering, have much greater mathematical content now than in the mid-twentieth century. All undergraduate majors in these areas have significant mathematical needs, as described in Section B. However, the formal mathematical requirements for majors in these areas have not grown in proportion to what is needed to pursue certain mathematically sophisticated sub-areas at an advanced level. Individuals with joint majors or dual majors in the mathematical sciences and a partner discipline are needed to meet these needs.

- There is a severe shortage of qualified teachers of secondary school mathematics. For example, in his 1999 article in the *Educational Researcher*, Richard Ingersoll reports that in 1995 one third of secondary mathematics teachers were teaching out of their field, having prepared to teach in another area. This shortage can be addressed only by graduating more mathematical sciences majors who are interested in secondary teaching.

- Many graduate mathematics departments have difficulty recruiting well-qualified graduates of U.S. colleges and universities to meet their need for teaching assistants and subsequently to fill positions in industry and universities requiring advanced degrees. Less than 50% of mathematics doctoral recipients from U. S. institutions are U. S. citizens[50] and greater numbers of well-prepared undergraduate majors with an interest in studying mathematics at an advanced level must be prepared to address this critical need.

- The fact that mathematics is a cornerstone of modern society implies that the study of mathematical sciences is important for all students, but it also implies that it is important that some leaders in all areas have the broader and deeper knowledge of mathematics conveyed by a degree in the mathematical sciences. Indeed, business, law, medicine and other professional schools seek mathematical sciences majors, and would welcome more.

While the need for graduates continues to increase, the number of majors has, as already noted, declined. (See Appendix 3 for an analysis of the data on numbers of majors.) At a time of increased need this decline is alarming at the national level; in some cases it also has very negative implications locally. If enrollments in advanced courses fall below threshold values at an institution, the availability of those courses decreases for students who want and need to take them.[51] Without a sufficient number of majors, it is difficult for a department to offer the range of courses and co-curricular experiences that best serve its students.

The population of potential majors has changed.

- Potential mathematical sciences majors—like all post-secondary students—are more diverse than they were even 30 years ago. English is not the primary language for many students. A very small number of students arrive in college and university classrooms well-prepared, highly interested in mathematics and intending doctoral study in mathematics. Most are less intrinsically interested in mathematics and lack confidence in their mathematical abilities; they may choose to major in mathematics because of its applicability in other disciplines or because it offers the promise of employment opportunities.

- Many potential majors begin their studies at two-year colleges. Over one third of recent bachelor's degree recipients in the mathematical sciences had taken courses at two-year colleges.[52]

---

[50] In 2001–2002 only 428 of the 962 doctoral recipients were U. S. citizens according to the Annual Survey of the Mathematical Sciences conducted by the AMS. www.ams.org/employment/surveyreports.html.

[51] See Table 4-3 in Appendix 4 for data on the declining availability of advanced courses. At an urban university that is a record-setting producer of mathematics graduates who are members of under-represented groups, there are sometimes too few students enrolled to offer differential equations.

[52] See Appendix 4.

- Not only are prospective mathematics majors more diverse in preparation and in career goals than twenty or thirty years ago, their goals are likely to change during their college years. Consequently it's not reasonable for departments and faculty to expect first and second year students to know what they want to study.

- In the 1960s, 5% of freshmen entering colleges and universities were interested in majoring in mathematics and 2% subsequently majored in mathematics, so it appeared departments were "filtering" prospective mathematics majors.[53] However, for the past twenty years, the percentage of entering freshmen intending to major in mathematics has been smaller than the percentages graduating with majors in the discipline, so departments appear to be "recruiting" majors (and many are succeeding).[54]

The recommendations given below for programs for the major are guided by the changing nature of the mathematical sciences, but also by this clear need for more mathematical sciences graduates in the face of the decline in the number of majors and the changing demographics of the student body.

Recommendation 1, that a department should understand the strengths and aspirations of its students and evaluate its courses and programs in light of its students and the resources of its institution, is particularly important for the mathematical sciences major. Admittedly, it would make life easier if these CUPM recommendations could include a list of courses describing the ideal mathematical sciences major. However, such a list is neither possible nor desirable. It is not possible because of the varying demographics and aspirations of students at diverse institutions nationally. It is not desirable because of the varied careers and fields in which mathematical sciences majors are needed and the capacities of different institutions to meet these different societal needs. Even at a single institution, providing a flexible major or a variety of tracks within the major can position the department to meet the diverse needs of its students most effectively.[55]

The recommendations for mathematical sciences majors include all those in Part I, which addressed all students, as well as the following specifically addressing students majoring in the mathematical sciences. The goals expressed in each recommendation are both desirable and attainable. Illustrative Resources contains examples that work toward implementation of the recommendations in a variety of settings.

C.1. Develop mathematical thinking and communication skills

Courses designed for mathematical sciences majors should ensure that students

- *Progress from a procedural/computational understanding of mathematics to a broad understanding encompassing logical reasoning, generalization, abstraction, and formal proof;*

- *Gain experience in careful analysis of data;*

- *Become skilled at conveying their mathematical knowledge in a variety of settings, both orally and in writing.*

Recommendation 2 states that every course should incorporate activities that will help all students progress in developing analytical, critical reasoning, problem-solving, and communication skills. This is particularly important for majors.

---

[53] *Models that Work*, page 4.

[54] In 2000, 1% of all bachelor's degrees were awarded in mathematics; in 1998 0.6% of entering freshmen intended to major in mathematics. See Appendix 3.

[55] See the preliminary report from Gilbert Strang at the Massachusetts Institute of Technology in the online *Illustrative Resources*, C.3.

Progress from a procedural/computational understanding of mathematics to a broad understanding. The ability to read and write mathematical proofs is one of the hallmarks of what is often described as mathematical maturity. Some of the pieces that go into this ability include careful attention to definitions, examination of the effects of modifying hypotheses, understanding of logical implication, and appreciation for how different aspects of a problem are related. Although each mathematics class will address the development of these abilities in its own way, some of the foundation for this kind of logical thinking must be laid in every course in which a prospective mathematics major might enroll, including calculus and discrete mathematics.

By the time they graduate, mathematical science majors should be expected to understand and write proofs. This does not imply that most of their coursework should consist of studying polished proofs of theorems. Neither is this the way mathematics is discovered/created nor is it the way to help most students master the underlying mathematical ideas or learn to write their own proofs. Faculty can best help students master the concept of proof by an incremental and broadly based approach; *every* advanced mathematics course can and should make its contribution to the development of students' ability to understand and construct mathematical proofs. Students should learn a variety of ways to determine the truth or falsity of conjectures. They need to learn to examine special cases, to look for counterexamples, and know how to start a proof and how to recognize when it is complete. Significant experience with proof can occur not just in traditional courses such as algebra, analysis, and geometry, but also, as indicated in Illustrative Resources, in courses as varied as cryptography, image compression, and linear models in statistics.

Gain experience in careful analysis of data. The analysis of data provides an opportunity for students to gain experience with the interplay between abstraction and context that is critical for the mathematical sciences major to master. Experience with data analysis is particularly important for majors entering the workforce directly after graduation, for students with interests in allied disciplines, and for students preparing to teach secondary mathematics. A variety of courses can contribute to this experience, including calculus, differential equations and mathematical modeling, as well as courses with a more explicit emphasis on data analysis. For example, students in a calculus course can fit logistic and exponential models to real data or encounter finding the best-fitting line for data as an optimization problem.

Convey mathematical knowledge, both orally and in writing. Careful reasoning and communication are closely linked. A student who clearly understands a careful argument is capable of describing the argument to others. In addition, a requirement that students describe an argument or write it down tests whether understanding has truly occurred. All courses should include demands for students to speak and write mathematics, and more advanced courses should include more extensive demands. Communicating mathematical ideas with understanding and clarity is not only evidence of comprehension, it is essential for learning and using mathematics after graduation, whether in the workforce or in a graduate program.

The 1991 CUPM report states, "Students who complete mathematics majors have often been viewed by industry, government, and academia as being well-prepared for jobs that require problem-solving and creative thinking abilities." The recommendations of the current *Guide* provide a basis to ensure that this reputation is upheld and enhanced.

C.2. Develop skill with a variety of technological tools

All majors should have experiences with a variety of technological tools, such as computer algebra systems, visualization software, statistical packages, and computer programming languages.

Recommendation 5 states that courses at all levels should: 1) incorporate activities that will help students learn to use technology as a tool for solving problems, and 2) make use of technology as an aid to understanding mathematical ideas. This recommendation and the discussion that followed it is applicable to courses designed for mathematics majors.

The first part of the recommendation states that students should be able to use technology as a tool. Learning to use technology effectively is an absolute necessity for students entering the job market immediately after receiving a bachelor's degree. Anyone doing technical work, including many in the teaching professions, will make extensive use of software; mathematical sciences majors need to prepare themselves by acquiring appropriately extensive experience with these tools.

The second part of the recommendation states that technology should be used as an aid to understanding mathematical ideas. Computer algebra systems, visualization software, and statistical packages can all be incorporated into courses in ways that facilitate exploration of concepts to a much greater extent than is possible with paper and pencil. These tools also enable students to focus on the big picture while working on complex problems. While technology is sometimes used in ways that obscure fundamental ideas and algorithms, this problem can be avoided by careful attention to design of assignments, construction of tests, and other aspects of pedagogy. On balance, the benefits of developing instructional techniques that use technology as an aid to understanding far outweigh the costs.

As described in Illustrative Resources, many departments are requiring majors to become proficient with graphing calculators, computer algebra systems, statistical packages, spread sheets, and programming languages during their first two years of study so that these tools can be used regularly and easily in large numbers of upper-level courses.

C.3. Provide a broad view of the mathematical sciences

All majors should have significant experience working with ideas representing the breadth of the mathematical sciences. In particular, students should see a number of contrasting but complementary points of view:

- *Continuous and discrete,*
- *Algebraic and geometric,*
- *Deterministic and stochastic,*
- *Theoretical and applied.*

Majors should understand that mathematics is an engaging field, rich in beauty, with powerful applications to other subjects, and contemporary open questions.

Every major should experience the breadth of mathematics. These contrasting points of view can often be encountered in the same course. Most programs include topics that could be described by the first of each of the dichotomies above, but many do not include sufficient emphasis on the second. Therefore, the following clarifications will focus on the second.

Continuous and discrete. Continuous mathematics (calculus, differential equations, analysis) typically is strongly represented in the mathematics major, with many departments requiring four or five courses. Indeed, calculus remains crucially important as the language of science and economics. All majors should study calculus, and many will be interested in areas that require extensive course work in this area. However, some or all of discrete mathematics, matrix algebra, probability and statistics, and other topics are even more important for many other students' career interests. Therefore the continuous and the discrete must both be present for all students. Allowing students to use discrete mathematics courses as prerequisites for higher-level mathematics courses can draw able and interested students into courses such as linear algebra, computational algebraic geometry, abstract algebra, and number theory, and increase the likelihood that they will decide to pursue a joint or double major with mathematics.

Algebraic and geometric. Algebra is typically represented in the mathematics major in many ways, beginning with the use of algebraic techniques in calculus and the study of linear algebra. Indeed, the program of every mathematics major should include linear algebra, and many students are well-served by continuing their study of algebra in courses such as advanced linear algebra, abstract algebra, coding theory, number theory, or linear models.

On the other hand, geometry is typically under-represented in the major. A number of contributors to *CUPM Discussion Papers about Mathematics and the Mathematical Sciences in 2010: What Should Students Know?*[56] call attention to the importance of geometry in the undergraduate curriculum. Herb Clemens points out that the "rush to calculus" in the high schools means students arrive in college with much weaker spatial skills, and yet there is no vehicle for remediating these deficiencies in the first two years. And at the advanced level, he observes, "There is plenty of justification coming from the directions of current research and applications in the very fields (analysis, applied math, probability and statistics, and even abstract algebra) which are the so-called competitors of geometry for time and attention." He also notes the particular needs of majors preparing to teach secondary mathematics, urging that "the fundamental course in geometry for [them should be] a serious course in two-dimensional geometry concentrating on non-trivial results in Euclidean geometry, but also with the three or four fundamental results of spherical and hyperbolic geometry." (pp. 23, 25) Roger Howe writes, "Geometry today is clearly the invalid of the college mathematics curriculum. In many institutions there is no regular offering in geometry, and in others, the main offering is a course specifically aimed at high school teachers.... It should not be so. Geometry was for centuries the heart of mathematics [and] ... still is a vital part. As represented in Lie theory, algebraic geometry, dynamical systems, Riemannian geometry and other research areas, it is a central part of the research enterprise." (p. 46)

Several tables in Appendix 4 corroborate the concern about geometry. Table 4-3 shows that from 1995 to 2000 the percentage of departments offering geometry courses fell from 69% to 56%. Also, Table 4-5 shows geometry enrollments falling along with the decline in total enrollments in advanced mathematics courses from 1985 to 1995. Geometry enrollments remained flat (at about half the 1970 enrollment) from 1995 to 2000, while total enrollment in advanced mathematics courses returned to 1985 levels.

Attention to geometric thinking should not be confined to geometry courses. It is also important in other courses, such as calculus and linear algebra. Illustrative Resources includes descriptions of a variety of geometry courses appropriate for majors with varying interests as well as of other courses—such as multivariable calculus, complex variables, linear models in statistics, and number theory—that emphasize geometric thinking and visualization and/or make effective use of graphics software to enhance understanding and as an aid to problem solving.

Deterministic and stochastic. Deterministic models are among the glorious achievements of mathematics, and every major should study them. Such models are currently well-represented in programs for majors in the mathematical sciences, as they should be.

But stochastic methods are also valuable and are typically less well-represented in the major program. Students should see examples of probabilistic methods in courses such as calculus and linear algebra as well as in the analysis of data. In particular, the *CUPM Guide 2004* supports the 1991 CUPM recommendation that every mathematical sciences major should study statistics or probability with an emphasis on data analysis for reasons including the following:

- Data analysis is crucial in many aspects of academic, professional and personal life;
- The job market for mathematical sciences majors lies heavily in fields that need people who can effectively draw conclusions from data;
- The emphasis on data analysis in the 2000 NCTM standards[57] and the growth of Advanced Placement statistics courses in secondary school make a study of statistics necessary for those preparing for secondary school teaching in mathematics.

---

[56]MAA Report, 2001.

[57]*Principles and Standards for School Mathematics,* National Council of Teachers of Mathematics, 2000. An electronic version is available at standards.nctm.org/info/about.htm.

The report of the statistics workshop in the Curriculum Foundations project argues that it is more important that mathematics majors study statistics or probability with an approach that is data-driven than one that is calculus-based, and CUPM agrees. The Guidelines of the American Statistical Association/ Mathematical Association of America Joint Committee on Undergraduate Statistics[58] specify that an introductory course should (1) emphasize statistical thinking: the importance of data production, the omnipresence of variability, the quantification and explanation of variability; (2) include more data and concepts, less theory, and fewer recipes; and (3) foster active learning. CUPM endorses the study of statistics for majors following an approach that satisfies the ASA/MAA guidelines. Such study also could serve as an alternative entry point for the major or minor.

Theoretical and applied. The theoretical ideas of mathematics must have an important place in the program of every major. Indeed, majors who have not seriously engaged rigorous theoretical mathematics will not fully understand the coherence, the beauty, and the power of the discipline.

However, recent advances in the mathematical sciences and in technology have implications for the balance between the theoretical and the applied in the undergraduate major. New applications of mathematics and new kinds of mathematics have become both important and possible to teach at the undergraduate level. A list of the mathematics used in different occupations appears on the careers webpage assembled by three of the professional societies in the mathematical sciences.[59] The list includes probability and statistics, ordinary and partial differential equations, Monte Carlo methods, optimization, control theory, game theory, discrete mathematics, splines, information theory, image compression and wavelets, computational geometry, and computational algebraic geometry. Applications in these areas are accessible to undergraduates, and selected topics should be included in the undergraduate program of all mathematical sciences majors. Illustrative Resources includes examples of successful undergraduate courses incorporating several of these topics.

Programs that emphasize discrete, geometric, stochastic, and applied topics along with the continuous, algebraic, deterministic and theoretical will enable students to better understand the emerging nature of mathematics. Majors also need experiences in which they grapple with the powerful applications to other subjects and encounter contemporary open questions.

C.4. Require study in depth

All majors should be required to

- ***Study a single area in depth, drawing on ideas and tools from previous coursework and making connections, by completing two related courses or a year-long sequence at the upper level;***
- ***Work on a senior-level project that requires them to analyze and create mathematical arguments and leads to a written and an oral report.***

The 1985 Association of American Colleges (AAC) report *Integrity in the College Curriculum* "views study in depth as a means to master complexity, to grasp coherence and to explore subtlety. The AAC goals for study in depth are framed by twin concerns for intellectual coherence intrinsic to the discipline and for development of students' capacity to make connections...."[60] This captures well the rationale for requiring work in depth of majors in the mathematical sciences. The question is, by what means is depth achieved?

---

[58] See www.amstat.org/education/Curriculum_Guidelines.html and *Heeding the Call for Change*, pp. 3–11.

[59] The careers page of the MAA, AMS and SIAM is at www.ams.org/careers/mathapps.html.

[60] MAA-AAC report, p. 189.

Study a single area in depth. To achieve depth in the major, the 1991 CUPM report specified a "two-course sequence in at least one important area of mathematics" with at least a calculus-level prerequisite for the first. Examples included probability and statistics, combinatorics and graph theory, two courses in numerical analysis, or real and complex analysis, as well as the more traditional two semester sequences in algebra or analysis.

These are effective means, but there are other combinations that can work, for instance a pair of courses in which neither is a prerequisite for the other, although both have a common prerequisite. In such a pairing, connections must be drawn at different levels of sophistication for students with different backgrounds. Examples of this kind of pairing include not only real and complex analysis (both with perhaps multivariable calculus as a prerequisite) but also two chosen from abstract algebra, coding theory, computational algebraic geometry, projective geometry (each with a linear algebra prerequisite), or the pair operations research and mathematical modeling (each with a linear algebra and multivariable calculus prerequisite). See Illustrative Resources for details of these and other examples.

Work on a senior-level project. Working on a project provides an experience of depth even when the project doesn't expand the area of study in the curricular sense. Although its focus may be quite narrow, a project requires the kind of integration and synthesis that are the desired consequences of the depth requirement. The kind of project stipulated in this recommendation requires the extended argument and analysis that are also part of what is meant by mathematical depth.

Further, a project for which a student must read, write and orally present significant mathematics develops the communication skills that are vital in the workplace and in future study. Such a project can ease the transition to the world beyond the classroom, where students must learn and use mathematics more independently than ever before and must communicate it clearly to a variety of audiences, including others less mathematically knowledgeable than themselves. Many employers are much more interested in an enthusiastic and lucid explanation of a personal mathematical project than in any list of courses or topics covered. A project also is excellent preparation for the intense independent learning that will be required for students continuing their studies with graduate work.

In some departments, it is feasible to require a capstone course for majors that includes such a significant project. Illustrative Resources contains examples of how such stand alone courses are conducted at a variety of institutions. But other means are also possible. A project can be embedded in a variety of advanced courses, or appended to a course for additional credit. Departments at institutions with January mini-terms can use them for this purpose. Summer research projects and internships also afford opportunities for projects. See Illustrative Resources for examples of projects of these varied kinds. Each department has to choose means appropriate to its students and resources, but ensuring that every major has this learning experience should be a priority, both for its value to the student and for its value to the department in assessing what its majors have learned.

C.5. Create interdisciplinary majors

Mathematicians should collaborate with colleagues in other disciplines to create tracks within the major or joint majors that cross disciplinary lines.

This is a restatement of part of Recommendation 4, placed here for emphasis. Also see the discussion of interdisciplinary majors in B.1. The greatest growth in the number of degrees awarded by departments of mathematics and statistics has been in areas outside the "classic" mathematics major (see Appendix 3). Based on the spring 2001 CUPM sample of mathematics departments, one third of degrees granted that year were to joint or double majors (see Appendix 4). The mathematical sciences major at the University of Michigan, the applied mathematics major at Brown University, and the joint majors in applied science at UCLA are effective programs, as are the joint majors with biological sciences, computer science, economics or business at a number of large and small institutions. See Illustrative Resources for details of these and other examples.

This recommendation highlights the importance of a department knowing its students and understanding the particular strengths of its broader institution. The Curriculum Foundations project found significant support within the partner disciplines for joint, interdisciplinary majors. Every college or university has its own opportunities for fruitful collaboration, and each mathematical sciences department should offer at least one major or one option within a major that encourages students to combine in-depth study of mathematical sciences with an allied field.

C.6. Encourage and nurture mathematical sciences majors

In order to recruit and retain majors and minors, mathematical sciences departments should

- *Put a high priority on effective and engaging teaching in introductory courses;*
- *Seek out prospective majors and encourage them to consider majoring in the mathematical sciences;*
- *Inform students about the careers open to mathematical sciences majors;*
- *Set up mentoring programs for current and potential majors, and offer training and support for any undergraduates working as tutors or graders;*
- *Assign every major a faculty advisor and ensure that advisors take an active role in meeting regularly with their advisees;*
- *Create a welcoming atmosphere and offer a co-curricular program of activities to encourage and support student interest in mathematics, including providing an informal space for majors to gather.*

Put a high priority on effective teaching in introductory courses. Departments with large numbers of majors have almost without exception made strong teaching in introductory courses a priority. Faculty in introductory courses have a significant influence on both recruitment and retention of majors and minors. Upper-level mathematics majors also can act as mentors to students in introductory courses, visiting classes to share their own experiences in subsequent courses and internships.

Seek out prospective majors. While recruitment and encouragement of mathematics majors can begin as early as middle school, these recommendations are restricted to secondary and post-secondary recruitment. Most colleges and universities sponsor special days when prospective students visit campus and meet with faculty in various departments. Mathematics faculty should be carefully assigned the task of meeting with these students—first impressions mean a lot.

The next place recruitment of mathematics majors can occur is within courses at the college level. As implied by Recommendation B.3 on prerequisites, there are multiple ways to begin the serious study of mathematics. Calculus is just one of them. Faculty should reach out to students in first year courses in discrete mathematics, statistics, number theory, and geometry and in survey courses, as well as calculus. Never underestimate the power of a faculty voice encouraging a student to consider a major in mathematics. Even if only a small percentage of these students go on to become majors, that fraction can represent a sizable increase in a program. A larger fraction may take additional courses in mathematics and add both numbers and vitality to advanced courses. Encouragement and support are particularly important for students from groups that do not traditionally major in mathematics. The websites of the MAA program Strengthening Under-represented Minority Mathematics Achievement (www.maa.org/summa) and of the Association for Women in Mathematics (www.awm-math.org) include links to a variety of resources.

Two year colleges play an important role in providing introductory courses in mathematics. They enroll 44% of undergraduates in the U.S., 46% of first time freshmen and 49% of undergraduates who identify themselves as members of racial or ethnic minorities.[61] Departments should cultivate relationships with nearby two-year colleges and seek out their students with interests in mathematics.

Students taking mathematics to satisfy the requirements for another major should be encouraged to consider a double or joint major with mathematics or at least a minor in mathematics. The growing presence of mathematics in traditionally non-mathematical disciplines such as biology is a strong selling point for students, but faculty need to make sure that students know of the pervasive uses of mathematics in those disciplines.

Inform students about careers in the mathematical sciences. Career information is vital. The Mathematical Sciences Careers web page (`www.ams.org/careers/`) can serve as an extremely valuable resource as can the MAA document *101 Careers in Mathematics*.[62] Departments also should make certain that advisors have access to relevant information from their own institution's careers office about alumni employment historically, as well as from recent alumni and their employers.

Most students will experience multiple career changes over their working lifetimes, so continuing professional education is increasingly important. There are new master's degree programs offered not only by mathematical sciences departments but also departments such as computer science, statistics, operations research, economics, business, and engineering. Some include BA-MS "articulation programs," connecting undergraduate mathematics to graduate programs in other disciplines. Awareness of these newer options places added responsibility on advisors. Departments should be sure that advisors are well informed about alumni experience in these and other graduate programs. (See the discussion of Recommendation D.3.)

Set up mentoring programs. Department faculty need to focus on the needs of individual students. Only a small minority of mathematics majors will pursue doctoral study in the subject (less than 10% at most institutions—see Appendix 4), and faculty must recognize the different interests of their majors. One successful department maintains a bulletin board (and web page) of brief reports from their graduates that describe their academic programs and current employment.

Working as tutors or graders can increase students' interest in continuing the study of mathematics as well as deepening their understanding of the content of the course they are assisting. Some departments invite students who performed well in first year courses to apply for these jobs. Departments need to offer appropriate training and support to assure that the tutorial or grading experience is rewarding both for the students receiving assistance and for those providing it.

Assign every major a faculty advisor who actively meets with advisees. Advising of majors throughout their undergraduate years is an important department responsibility. Unlike an earlier, simpler day when all mathematics majors took the same sequence of courses with only a few electives in the senior year, the typical undergraduate mathematical sciences department today requires students to make substantial curricular choices. Advisors should carefully monitor each advisee's academic progress and changing goals and work with the student to explore the many intellectual and career options available to mathematics majors. For some students, achieving the best choice of courses may necessitate coordination between the major advisor and faculty in another department.

Advisors also should pay particular attention to the need to retain capable undergraduates in the mathematical sciences pipeline, with special emphasis on the needs of under-represented groups. When a

---

[61] "First Steps: The Role of the Two-Year College in the Preparation of Mathematics-Intensive Majors," by Susan S. Wood, in *CUPM Discussion Papers about Mathematics and the Mathematical Sciences in 2010: What Should Students Know?*, MAA Reports, 2001, p.101.

[62] *101 Careers in Mathematics*, 2nd edition, edited by Andrew Sterrett, Classroom Resource Materials, MAA, 2003.

department offers a choice of several tracks within the major, advisors have the added responsibility of providing students with ample information even when students do not ask many questions. This individualized approach to advising requires that no advisor be assigned too many advisees.[63]

Create a welcoming atmosphere and offer a co-curricular program of activities to encourage and support student interest in mathematics. Departments should provide space for informal student contact. Students who develop good working relationships with peers are more likely to succeed in their mathematics courses and to sustain interest and motivation. A student lounge equipped with work tables, writing boards and (comfortable) chairs should be available for student/student as well as student/faculty contact. Shelves containing current student-friendly mathematical journals as well as selected mathematics books would add to the atmosphere of such a room.

Departments should also offer opportunities for mathematics majors to interact with one another through a campus organization. This might be a mathematics club, a student chapter of the MAA or a chapter of Pi Mu Epsilon. Such an organization can serve as a catalyst for a number of retention activities — mentoring programs, attendance at local or national meetings, and sponsorship of movies and presentations. Such an organization also gives students a sense of ownership in their major.

D. Mathematical sciences majors with specific career goals

D.1. Majors preparing to be secondary school (9–12) teachers

In addition to acquiring the skills developed in programs for K–8 teachers, mathematical sciences majors preparing to teach secondary mathematics should

- *Learn to make appropriate connections between the advanced mathematics they are learning and the secondary mathematics they will be teaching. They should be helped to reach this understanding in courses throughout the curriculum and through a senior-level experience that makes these connections explicit.*

- *Fulfill the requirements for a mathematics major by including topics from abstract algebra and number theory, analysis (advanced calculus or real analysis), discrete mathematics, geometry, and statistics and probability with an emphasis on data analysis;*

- *Learn about the history of mathematics and its applications, including recent work;*

- *Experience many forms of mathematical modeling and a variety of technological tools, including graphing calculators and geometry software.*

The teacher preparation recommendations of this *Guide* have been informed by *The Mathematical Education of Teachers (MET),*[64] a recent CBMS report that presents detailed and carefully considered guidelines concerning the education of future teachers of mathematics. Mathematics faculty and departments are advised to study MET in its entirety.[65] Prospective teachers of secondary school mathematics

---

[63] This recommendation and discussion are adapted from the 1991 CUPM report, which reappears as Appendix E of *MAA Guidelines for Programs and Departments in Undergraduate Mathematical Sciences,* and from the recommendations in the *MAA Guidelines for Programs.*

[64] *The Mathematical Education of Teachers*, volume 11 of the Issues in Mathematics Education series of the Conference Board of the Mathematical Sciences, AMS and MAA, 2001, available at www.cbmsweb.org.

[65] The *MET* authors write, "This report is not aligned with a particular school mathematics curriculum, although it is consistent with the National Council of Teachers of Mathematics' *Principles and Standards for School Mathematics* as well as other recent national reports on school mathematics."

should complete a major in the mathematical sciences.[66] Recommendation D.1 is thus a companion to the recommendations for all mathematical sciences majors in Section C. It does not, however, specifically address course work in education.

It bears repeating that the postsecondary education of future teachers of mathematics is an important responsibility of mathematicians. The need for qualified teachers of mathematics is great, and students potentially interested in teaching mathematics should be encouraged and supported, as well as prepared to be effective teachers. David Sánchez argues that mathematics faculty concerned about the quality of mathematics teaching in the schools should work to ensure

> ... that the future teachers they produce have the best possible training and strong, conceptual understandings of mathematics. Hopefully, those teachers will be the leaders of reform by working within the educational system through mathematics and leadership workshops, becoming master teachers, and exemplifying the best in content mastery and pedagogy. Their undergraduate training will be a cornerstone of that effort.[67]

As its authors acknowledge, the *MET* recommendations "outline a challenging agenda for the education of future secondary school mathematics teachers," and each department will have to adjust its program, in ways that fit its mission and resources, to meet the challenge. Fortunately, these adjustments are likely to serve *most* majors well.

Connect advanced and secondary mathematics. The assumption that the traditional curriculum for a mathematics major is adequate preparation for students preparing to teach secondary school is simply incorrect. It is not enough for secondary mathematics teachers to have an understanding of advanced mathematics—they must also be able to connect their advanced coursework to the material they will teach.[68] For example, secondary school teachers should

> ... be able to answer the following typical high school students' questions, in ways that are both mathematically sound and also accessible and compelling to a 15 year-old:
> - Why does a negative times a negative equal a positive?
> - Why do I switch the direction of the less than symbol when I multiply both sides by a negative number?
> - In every triangle that I tried in *Sketchpad,* the angles add up to 180°. I don't need to do a proof, do I?
> - I am not convinced that .99999... = 1.
> - How do I know parallel lines never intersect?
> - I think that the number 1 has three different square roots: 1, –1, and .99999999.
> - I am sure that .99999999 is a square root of 1 because when I multiply it by itself on my calculator I get 1.00000000.[69]

---

[66] As in Section C, this includes programs designed for prospective secondary mathematics teachers, whether they are called "mathematics" or "mathematics education" programs.

[67] "The Mathematics Major Overview," by D. Sánchez, in *CUPM Discussion Papers on Mathematics and the Mathematical sciences in 2010: What Should Students Know?*, MAA Reports 2001, p. 82.

[68] This point was made in 1996 in the presentation "On the education of mathematics teachers," by H. Wu at the Mathematical Sciences Research Institute. A later adaptation is Wu's "On the Education of Mathematics Majors," in *Contemporary Issues in Mathematics Education*, ed. by E. Gavosta, S.G. Krantz and W.G. McCallum, MSRI Publications, Volume 36, Cambridge University Press, 1999, pp. 9–23. Both papers (and related material) are available at www.math.berkeley.edu/~wu.

[69] "The Mathematical Education of Prospective Teachers of Secondary School Mathematics: Old Assumptions, New Challenges," by J. Ferrini-Mundy and B. Findell, in *CUPM Discussion Papers about Mathematics and the Mathematical Sciences in 2010: What Should Students Know?*, MAA Reports, 2001, p. 34.

Highlighting these mathematical connections in upper-level courses is thus an important component of the education of future secondary mathematics teachers. The *MET* and other resources include many suggestions for modifying traditional mathematics courses to help prospective teachers make these connections. For example, a linear algebra course can include study of the rigid motions of the plane and relate this topic to Euclidean geometry. Assignments in algebra or number theory "might ask for the use of unique factorization and the Euclidean algorithm to justify familiar procedures for finding common multiples and common divisors of integers and polynomials" or "ask for each step in the solution of a linear or quadratic equation to be justified by a field property."[70] Attention to the connections between advanced material and topics studied in secondary school would be valuable for all majors. (See Illustrative Resources for additional ideas.)

Departments should require prospective secondary mathematics teachers, like all mathematical sciences majors, to complete a senior-level intensive project (Recommendation C.4). This project should help future teachers explore further the relationship between advanced mathematics and the mathematics they will be teaching.

Include specific topics in mathematics. Although many different choices of topics can provide strong preparation for mathematical sciences majors, prospective teachers need to study advanced material related to the topics they will likely teach.

- ***Abstract algebra and number theory***. "The algebra of polynomial and rational expressions, equations and inequalities has long been the core of high school mathematics."[71] Therefore, students preparing to teach secondary mathematics need a solid understanding of number systems, structures underlying rules for operations, and the use of algebra to model and solve real world problems. Indeed, all majors, not just prospective teachers, should know *why* our number systems behave as they do. They should be able to explain specific facts like $1/(1/5) = 5$. They should be able to explain the relationship between the algorithms of arithmetic for numbers in decimal representation and the arithmetic of polynomial algebra. An abstract algebra course for prospective teachers should include the study of rings and fields with explicit attention to the rational numbers—not just as an abstract set of equivalence classes of ordered pairs of integers, but as the rational numbers appear in the secondary school curriculum. The twin problems of integer factorization and primality testing can be used to explore the structure of the integers. A traditional 'groups, rings, fields' course can be adapted, with appropriate elaborations and emphases, but there are other possibilities too. For example, an algebra course might be organized around "the solutions of the three classical construction problems and the explanation of why the roots of some polynomials of degree greater than or equal to five cannot be extracted from the coefficients by use of radicals."[72]

- ***Analysis***. Analysis provides "a rigorous foundation for future teaching about functions and calculus. Informal notions about Euclidean space, functions and calculus that undergraduates have used for several years can be given sound formal definitions."[73] Although mathematics majors have a lot of exposure to functions, many develop only a formulaic approach to their use. Future secondary teachers should develop an understanding of the main characteristics of the basic functions in school mathematics. For example, they should know that the key features of polynomial functions are their zeroes

---

[70] *The Mathematical Education of Teachers*, page 40.

[71] *The Mathematical Education of Teachers*, page 40.

[72] "On the Education of Mathematics Majors," H. Wu.

[73] *The Mathematical Education of Teachers*, page 134.

and the shapes of their graphs, that the exponential and logarithm functions owe their importance to modeling natural growth and decay as well as converting addition to multiplication and *vice versa*, and that the trigonometric functions are needed to model all periodic phenomena. Also, in practice school mathematics deals primarily with the rational numbers, not the real numbers, but the connection between the two number systems is rarely made clear. One goal of the study of analysis by prospective teachers should be to address this "missing link" to prepare them to handle this implicit but important issue in the classroom.[74]

- **Discrete mathematics.** The *MET* recommends that the program of prospective secondary mathematics teachers include work in discrete structures and their applications, design and analysis of algorithms, and the use of computer programming to solve problems. More specifically, the *MET* states that prospective secondary mathematics teachers should be exposed to graphs, trees and networks, enumerative combinatorics, finite difference equations, iteration and recursion, and models for social decision-making. Future teachers can encounter these topics in a variety of courses as well as in a discrete mathematics course. For example, iteration and recursion might appear in a calculus or differential equations course; finite difference equations and models for decision-making might be in a modeling course; graphs, trees and networks might be in a "bridge" course designed to help students make the transition from introductory to advanced courses.

- **Geometry.** Teachers of secondary school geometry need to understand the concepts of Euclidean geometry. In particular, they need to have well-developed geometric intuition—built up with hands-on experiences such as ruler and compass constructions, making three-dimensional models of polyhedra and curved surfaces, and/or creating tilings of the plane with polygons —so they can help their future students acquire it. Teachers should also know the logical structure of the subject. For example, they should know the relationship between what assumptions are made about parallelism of lines in a geometry and proving a theorem about the sum of the angles of a triangle. Teachers should be able "to use dynamic drawing tools to conduct geometric investigations emphasizing visualization, pattern recognition, conjecturing, and proof."[75] Teachers need to have facility with proof themselves in order to be able to lead their students from intuition and the drawing of pictures to logical reasoning. Prospective teachers should become fluent with "proofs by local axiomatics": starting with a fixed collection of (geometric) facts and deducing from them something interesting, and they should have some experience with the more subtle and difficult deductions from axioms.[76] Teachers should also be acquainted with recent developments in geometry and its uses.

- **Statistics and probability with an emphasis on data analysis.** Data analysis and probability form one of the core strands in the current K–12 curriculum. "Because statistics is first and foremost about using data to inform thinking about real-world situations, it is critical that prospective teachers have realistic problem-solving experiences with statistics."[77] In addition, the *MET* recommends that secondary mathematics teachers should have experience with exploring data, planning a study, anticipating patterns, using statistical inference, and applying appropriate technologies. A course meeting the ASA/MAA guidelines (as described in the discussion of Recommendation C.3 and elaborated in Illustrative Resources) would serve prospective teachers well.

---

[74]Comment to CUPM by H.Wu.

[75]*The Mathematical Education of Teachers*, page 41.

[76]Comment to CUPM by H.Wu.

[77]*The Mathematical Education of Teachers*, page 44.

Learn about the history of mathematics and its applications. Prospective secondary mathematics teachers need to know the history of the subjects they will teach. It gives them a better appreciation for the struggle that goes into mathematical advances. It enables them to identify conceptual difficulties and to see how they were overcome. And it enriches their own understanding of the mathematics they will teach and the role it has played in human history. This history of mathematics needs to include modern developments, as in Recommendation 3.

> Because of its enormous practical value, mathematics is frequently taught as a collection of technical skills that are applicable to specific tasks and often presented without reference to the intellectual struggles that led to contemporary understanding.... many non-European cultures have made sophisticated and significant contributions to mathematics.... [F]uture secondary mathematics teachers will be well-served by deeper knowledge of the historical and cultural roots of mathematical ideas and practices.[78]

While some departments may offer courses in the history of mathematics, others will explore these themes in a variety of classes, including perhaps the senior-level intensive project (C.4).

Experience mathematical modeling and a variety of technological tools. Prospective teachers should experience many forms of mathematical modeling, such as differential and difference equations, linear statistical models, probability models, linear programming, game theory, and graph theory. Work with a model should include attention to assessing how well models fit the observed facts. The *MET* discusses modeling real phenomena and using models to draw conclusions, largely in the context of stochastic models. The use of stochastic models is indeed important. Experience with deterministic models is also important for prospective secondary school teachers. Such models can illuminate and motivate the algebra and geometry in the secondary mathematics curriculum and help secondary students see and appreciate the broad usefulness of mathematics in the world. In addition, modeling is an area in which the importance of stating one's assumptions clearly comes to the front: it reinforces the point that clear thinking and precision are crucial for problem solving as well as deductive reasoning. As in B.2, learning to create, solve and interpret a variety of discrete and continuous models is important for prospective teachers.

Not only should prospective teachers use technology to build their own understanding of mathematical ideas and as a tool to solve problems in their own courses, as in Recommendations 5 and C.2, they also have to be prepared to use these and other tools appropriately and effectively as teachers. The discussion above of geometry and data analysis alluded to the value of technology for learning these topics and solving meaningful problems. The study of families of functions is found in many middle and secondary school curricula, often in a modeling context. The graphing and table-generating features of calculators or spreadsheets can be used effectively to study such functions. As recommended in the Curriculum Foundations workshop on teacher preparation, "Future teachers should be able to use tools such as tiles, cubes, spheres, rulers, compasses, and protractors to deepen their understanding of the mathematics they will have to teach, including 2-D and 3-D geometry and measurement. The use of electronic drawing tools such as the Geometer's Sketchpad and Cabri is also recommended for use in geometry courses for both elementary and secondary teachers. The dynamic capabilities of these tools allow both students and teachers to test conjectures relatively easily."[79]

---

[78] *The Mathematical Education of Teachers*, p. 142.

[79] *The Curriculum Foundations Project: Voices of the Partner Disciplines*, p. 150.

D.2. Majors preparing for the nonacademic workforce

In addition to the general recommendations for majors, programs for students preparing to enter the nonacademic workforce should include

- *A programming course, at least one data-oriented statistics course past the introductory level, and coursework in an appropriate cognate area; and*

- *A project involving contemporary applications of mathematics or an internship in a related work area.*

More than 90% of mathematical sciences majors go directly into the workforce after graduation (see Appendix 4). Many of the recommendations in Part I for all mathematics students and in Section C for all mathematical sciences majors are of great importance for students entering the work force. The thinking skills described in Recommendations 2 and C.1 are vital. For example, the website www.careers-in-business.com lists the skills and talents required for work as a consultant, rating them in importance from medium (initiative, computer skills) to high (people skills, teamwork, creativity, ability to synthesize) to *extremely* high (just one: analytical skills). The Society for Industrial and Applied Mathematics (SIAM) lists as one of the most important traits of an effective employee outside academia "skill in formulating, modeling and solving problems" (www.siam.org/mii). A participant in a January 2000 CUPM focus group representing a variety of employers in business, government, and industry said his consulting firm "likes to hire math people" for the way they think, for their ability to formulate problems mathematically, and for their modeling skills.

Also mentioned on these two web pages and in virtually every discussion of preparation for the workforce is the importance of oral and written communication skills, as well as the ability to work well as a member of a team. Experience with a project of the kind recommended in C.4 is also highly valued, especially if the project requires working as a member of a team.

Programming, statistics, and cognate coursework. The 1993 NSF survey of recent graduates found nearly half of mathematical sciences bachelor's degree recipients in nonacademic positions spent substantial time working with computer applications. This is certainly no less true today. Programming skills enable a user to take advantage of the full power of many software packages and are a valuable selling point with potential employers. In "Business View on Math in 2010," Patrick McCray writes, "When evaluating a resume to decide whether or not to extend an invitation for a job interview to an applicant, ... I look at an inventory of skills, such as fluency with programming languages, computer environments.... Applicants who do not have the potential to be productive within two weeks of being hired are not even considered for an interview." [80]

The view of programming as consisting only of if-then, do-while, and a few other structures is several decades behind the current state of the art. Object-oriented languages such as *C* were a jump up in abstraction and complexity from *FORTRAN*, and *Pascal*, *Java*, and *.NET* are another jump up. If a person needs to learn a programming language in the future, the best basis is to know one of the state-of-the-art languages of today.

In the list of mathematics used in different occupations on the careers webpage assembled by three of the professional societies in the mathematical societies,[81] statistics and probability appear repeatedly. The 1997 and 1993 national surveys of recent recipients of bachelor's degrees in mathematical sciences find large numbers of students employed in sectors where data analysis and probabilistic modeling are impor-

---

[80] Paper by P.D. McCray of Pharmacia Corporation in *CUPM Discussion Papers about Mathematics and the Mathematical Sciences in 2010: What Should Students Know?*, MAA Report, 2001, p. 75.

[81] The careers page of the MAA, AMS, and SIAM is at www.ams.org/careers/mathapps.html.

tant. (Also, see the ASA web page at `www.amstat.org/careers/`.) Courses in statistics and probability are valuable assets to students seeking to enter the workforce after graduation, especially if these courses have involved work with substantial data sets and have required reports explaining and interpreting the results of the data analysis. Having cognate coursework that makes significant use of statistics or modeling further enhances a student's chances of finding employment.

Numerous institutions have developed combinations of courses that give majors access to a wide range of interesting and well-paid jobs. In addition, many courses beyond those listed above are valuable to students seeking non-teaching jobs after graduation (see Illustrative Resources).

Projects and Internships. Working in business, industry, or government helps students make good decisions about the kind of work they might enjoy and do well. Sometimes an internship can even open the door to specific employment opportunities. Internships teach students valuable lessons about how the knowledge and skills they gain in school can be deployed in the workplace. And, internships look good to prospective employers. McCray writes, "My advice to prospective applicants: While a student, get work experience in the same line of business in which you wish to seek employment after graduation…It is easier to gauge how a person will fit into a specific work environment if that person has already taken advantage of opportunities to explore similar ones."[82]

Departments often need the assistance of their institution's careers office to help students find good internships. Even institutions far from urban settings should have access to data bases of opportunities for undergraduates. Departments need to work cooperatively with careers offices to build resources and to take maximum advantage of resources that exist. Students also can help each other by reporting on their internship experiences at mathematics club meetings, brown bag seminars, or other informal sessions (see Illustrative Resources for examples).

A project based on an applied problem of contemporary interest is also valuable. Analyzing the problem, choosing appropriate mathematical or statistical tools, collecting relevant data, writing a clear description of the analysis and solution of the problem, and interpreting the results for a non-mathematical audience all build and reinforce skills needed in the workplace. Collaboration with colleagues in other departments and, where possible, with employers in the area can strengthen the project and add to the authenticity of the experience.

D.3. Majors preparing for post-baccalaureate study in the mathematical sciences and allied disciplines

Mathematical sciences departments should ensure that
- *A core set of faculty members are familiar with the master's, doctoral and professional programs open to mathematical sciences majors, the employment opportunities to which they can lead, and the realities of preparing for them;*
- *Majors intending to pursue doctoral work in the mathematical sciences are aware of the advanced mathematics courses and the degree of mastery of this mathematics that will be required for admission to universities to which they might apply. Departments that cannot provide this coursework or prepare their students for this degree of mastery should direct students to programs that can supplement their own offerings.*

There are many graduate and professional degree programs that actively solicit mathematics majors. In addition to the traditional master's and doctoral programs in pure and applied mathematics, many disci-

---

[82] *CUPM Discussion Papers about Mathematics and the Mathematical Sciences in 2010: What Should Students Know*, page 75.

plines with rich mathematical content offer professional master's degrees. These include statistics, financial mathematics, operations research and industrial engineering, biomedical imaging, molecular structure and pharmaceutical sciences, natural resource management, and cryptography. There also are doctoral programs in many of these fields. Medical schools, law schools and business schools also recognize mathematics as strong undergraduate preparation for their programs. Departments need to be aware of the varied opportunities for further study that are available to their students.

The professional (terminal) master's degree in the mathematical sciences is gaining both visibility and importance. As reported in *SIAM News*, there is "a new class of professional science master's programs, many offering rich interaction with business, industry, and government."[83] The Alfred P. Sloan Foundation and the William M. Keck Foundation have supported the establishment of many new multidisciplinary master's programs with strong mathematical components, including applied financial mathematics, applied statistics, financial mathematics, industrial mathematics, mathematical sciences, mathematics for entrepreneurship, quantitative computational finance, and statistics for entrepreneurship.[84] The Society for Industrial and Applied Mathematics and the Commission on Professionals in Science and Technology are tracking the performance of these programs. More information will be available on their websites.

A core set of faculty members should be identified to advise students interested in the growing number of cross-disciplinary graduate programs that combine mathematics with another subject. Students in these programs will need fluency in both mathematics and the partner subject. They will need to be able to handle both precision and approximation, modeling and analysis, problem solving and careful reasoning, communication with computers and with people. Because these fields are evolving rapidly, a strong intellectual foundation is essential to facilitate career-long learning.

Departments should be able to prepare students for the wide variety of post-baccalaureate programs within the mathematical sciences. These include mathematics, applied mathematics, statistics, applied statistics, actuarial science, operations research, and mathematics education. Requirements vary, but facility with the tools of calculus, statistics, differential equations, and linear algebra is basic to all of these programs. Facility implies more than having taken a course in that topic. It means that the student understands the ideas that lie behind the tools and is adept at solving problems that require those tools. It also means that the student knows when and how a particular mathematical idea is likely to be most useful.

The requirement that all mathematics majors should have some familiarity with at least one programming language applies as much to those intending to pursue graduate work as to those headed directly for employment outside academia. This might be programming in *Maple, Mathematica, MatLab, Java* or *C++*; programming in *SPlus, R,* or *SAS* is valuable for students planning graduate study in statistics. Departments should know which languages and professional software will be most useful for students heading into the various graduate programs.

Because these graduate programs and their requirements are so varied, it is important that certain faculty be designated to keep current on the requirements for different programs. These faculty can serve as liaisons with programs that have the potential to attract graduates from their institution. They should arrange to bring in representatives from different graduate programs including some of their own alumni who can talk about a variety of graduate experiences

For majors intending to pursue doctoral work in the mathematical sciences, faculty must communicate the fact that facility with the tools of calculus, differential equations, and abstract linear algebra is funda-

---

[83]"SIAM Survey to Target Professional Master's Programs," *SIAM News*, Volume 36, Number 1, January/February 2003.

[84]A website with links to all these programs can be found at www.sciencemasters.com.

mental, as is some knowledge of programming. It is not possible to prescribe exactly what preparation students will need; much depends on the program they wish to enter. For a doctoral program in statistics (and for some master's programs), an analysis course is essential; students taking analysis as undergraduates have an advantage for admittance to doctoral programs in statistics. For both master's and PhD statistics programs, a strong cognate area (i.e., depth through several undergraduate courses or a second major) is valuable (e.g., biology, economics, physical science).

Doctoral programs in pure mathematics almost always require their students to take graduate courses in analysis and algebra and to pass the demanding qualifying exams based on these courses. Departments that prepare students for doctoral work in mathematics should ensure that their students arrive in graduate school prepared to take such courses.

Doctoral programs in pure mathematics assume that incoming students are already familiar with the fundamental definitions, concepts, and theorems of analysis and algebra. They often presuppose additional knowledge of topics in complex analysis, geometry, and topology. There may be flexibility in that a student may be able to rely instead on knowledge of numerical analysis, logic, or another mathematical subject, but algebra and analysis are almost always central to the first year of graduate study. For example, the departmental website at the University of Illinois at Urbana-Champaign has a list of the basic topics with which they assume familiarity. For real analysis and abstract algebra, these consist of:

> ***Real analysis:*** *Completeness properties of the real number system; basic topological properties of n-dimensional space; convergence of numerical sequences and series of functions; properties of continuous functions; and basic theorems concerning differentiation and Riemann integration.*

> ***Abstract Algebra:*** *Modular arithmetic, permutations, group theory through the isomorphism theorems, ring theory through the notions of prime and maximal ideals; additional topics such as unique factorization domains and classification of groups of small order.*[85]

Some doctoral programs require additional topics in real analysis and abstract algebra or specify that preparation for their programs requires an undergraduate analysis class at the level of Rudin's *Principles of Mathematical Analysis* and an undergraduate algebra class at the level of Herstein's *Topics in Algebra*, while others will allow entering students to start with a preparatory course at the level of Rudin or Herstein.

Some programs have a wide variety of qualifying topics and students may have some choice of which qualifying examinations to take. In addition, some programs permit the student without the required background to catch up. Department advisors should advise students accordingly, including realistic appraisals of how much preparatory work will be needed before the student is ready to take required graduate courses.

Preparing students for doctoral work in mathematics requires more than exposing them to topics in real analysis and abstract algebra. Students need to be able to read and critique proofs. This skill includes the ability to determine how assumptions are used and to find counterexamples when any of the hypotheses are weakened. Students need to be familiar with the common techniques employed to prove results in analysis and in algebra, and they should be able to use these techniques to prove theorems they have not seen before. Above all, students heading into doctoral programs in mathematics need to be able to read with understanding the textbooks they will encounter written in the languages of analysis and algebra. This is a skill that departments should ensure that their students learn.

Participation in an REU[86] program is not a replacement for subject matter courses, but it can be good

---

[85] www.math.uiuc.edu/GraduateProgram/intro.html#Background. This should not be interpreted as an endorsement of these topics as the proper undergraduate preparation for a doctoral program in mathematics. It is simply an example of what one respected graduate program expects.

[86] NSF Research Experiences for Undergraduates.

supplemental preparation that introduces students to the excitement and hard work involved in doing mathematical research. There are other summer programs aimed at students considering doctoral study in mathematics and also semester-long programs for visiting students; see Illustrative Resources for details. Departments should be aware of these programs and direct promising students toward them.

Departments that cannot afford to provide the full preparation needed for the doctoral program to which a student aspires should consider providing options such as reading courses or cross-registration for courses at a nearby institution. They might recommend a transitional fifth year of study or transfer to another institution.

For some students, a master's degree in mathematics is an attractive option that permits them to "try out" graduate school as well as improve their preparation for doctoral study if they choose to continue. However, funding opportunities may be more limited for master's programs than for doctoral programs.

Illustrative Resources for *CUPM Guide 2004*

online at `www.maa.org/cupm/`

This online document describes a variety of experiences and resources associated with the *CUPM Guide2004*, following the organization in Parts I and II. As a major part of the development of this document, CUPM made a broad request for reports on experiences from individuals who had in recent years implemented various ideas discussed in earlier drafts of the *CUPM Guide 2004*, including specific requests to departments with large numbers of majors and/or recent NSF or FIPSE awards for curricular projects. (See Appendix 1 for additional details on the gathering of information.) There was a large response to the CUPM request, and thanks are due to the many mathematics faculty who gave generously of their time and wisdom. The contributed examples, experiences and resources were instrumental in developing this document, and they provided evidence that the CUPM recommendations are indeed feasible. References to many of these contributions are made in the Illustrative Resources

---

The experiences described or referred to in the Illustrative Resources for *CUPM Guide 2004* are far from exhaustive. They were not endorsed by CUPM, nor is there an implication that they represent better practice than other implementations not included there. Rather they are offered as experiences and ideas that appear interesting and may serve as a starting point for those considering enhancement of components of their department's programs.

CUPM invites all members of the mathematical sciences community to report on additional resources they have successfully used or developed to work toward the recommendations in *CUPM Guide 2004*. They should be submitted to `cupm@maa.org`. The Illustrative Resources will be updated and expanded periodically.

Most entries in the Illustrative Resources include a link to a web address that describes a site containing further material or provides a means of contacting a person or department for additional information. An annotated bibliography with additional reference information is at the end of the Illustrative Resources.

The following pages give an indication of what the Illustrative Resources contains.

CONTENTS OF ILLUSTRATIVE RESOURCES
(September 2003)

Part I. Recommendations for Departments, Programs, and all Courses in the Mathematical Sciences

1. Understand the student population and evaluate courses and programs

Understanding Entering Students via Placement Exams
Understanding Entering Students via Mathematical Autobiography
Advising Students—Understanding their Goals
Understanding and Supporting Strong Students
Efforts to Assess Undergraduate Mathematics Programs
Supporting Faculty and Department Efforts in Assessment of Undergraduate Mathematics
Assessment Tools in the Classroom
Gathering Information about Students and Alumni to Improve Programs

2. Develop mathematical thinking and communications skills

Inquiry-Based Learning
Research on Student Difficulties with Reasoning and Problem Solving
Reasoning and Logically Working to Conclusions
Problem-Based Learning and Unstructured Problems
Strategies for Problem Solving
Problem Posing
Efforts to Improve Student Reading
Developing Students' Ability to Read Mathematical Writing
Student Classroom Discussions
Getting Started—Writing and Expressing Mathematics
Evaluating and Assessing Students' Skills in Communicating Mathematics
Rubric for Grading Mathematical Writing Assignments
Additional Resources

3. Communicate the breadth and interconnections

Key Ideas and Concepts
Multiple Approaches/Perspectives
Examples and Applications
Contemporary Topics
Enhance Perception of Vitality and Importance of Mathematics
Additional Resources

4. Promote Interdisciplinary Cooperation

Connecting with other Disciplines within a Mathematics Course
Developing Interdisciplinary Courses
Curriculum of Interdisciplinary Projects
Developing Interdisciplinary Programs
Additional Resources

5. Use computer technology to support problem solving and understanding

Demonstrations for Understanding and Visualization
Exploration, Conjecture and Proof
Using Computer Language
Integrating Technology Tools
Introductory Mathematics with Technology
Additional Resources

6. Provide faculty support for curricular and instructional improvement

Teaching and Learning
Faculty and Professional Development Programs
Successful Practices at Specific Institutions
Additional Resources

Part II. Additional Recommendations Concerning Specific Student Audiences

A. Students taking general education or introductory collegiate courses in the mathematical sciences

A.1. Offer suitable courses

Offering Engaging Courses
Engaging Students in Project Work
Quantitative Literacy
Developing Mathematical and Quantitative Literacy Across the Curriculum
Offering Choices to Satisfy a General Mathematics Requirement
Examples of Introductory Course Syllabi
Support for Faculty Teaching Developmental Mathematics

A.2. Examine the Effectiveness of College Algebra

Refocusing College Algebra
College Algebra — New Approaches

A.3. Ensure the effectiveness of introductory courses

Pre-Calculus—New Approaches
Integrating Pre-Calculus and Calculus

B. Students majoring in partner disciplines & prospective teachers

B.1. Promote interdisciplinary collaboration

Strengthening Mathematics Courses to Support Future STEM Study
Course Pairings
(See C.5 for Interdisciplinary Majors)

B.2. Develop mathematical thinking and communication

Improving Students' Abilities to Think about and Do Mathematics
Research on Teaching and Learning
Programs to Promote Mathematical Insight into All Disciplines
Writing in Introductory and Service Courses

B.3. Critically Examine Course Prerequisites

Prerequisites

Including three-dimensional topics in the first year.

B.4. Pre-service elementary (K–4) and middle-school (5–8) teachers

Guidance to Colleges and Universities

Programs for Elementary Teachers

Programs for Middle School Teachers

Research to Practice—Elementary and Middle School Teachers

Programs for Mathematicians Teaching Future Teachers

C. Students majoring in the mathematical sciences

Examples of Effective Majors

Examples of Programs at Schools with a Large Number of Mathematics Majors

Additional Resources

C.1. Develop mathematical thinking and communication skills

Reasoning and Problem-Solving

Workshops for Mathematics Majors

Inquiry-Guided Learning

Learning through Proof

Developing Ideas of Formal Proof

Writing, Reading and Exploring Proofs

Gain Experience in Careful Analysis of Data

For Mathematics Majors—Reading, Writing and Expressing Mathematics

Evaluating and Assessing Majors' Skills in Communicating Mathematics

Draft Outline of Assessment Plan for Sample Learning Outcome

C.2. Develop skill with a variety of technological tools

Technology and Interactive Learning

Visualization

Linear Algebra—Computer/Technology-Based Texts and Courses

Calculus and Differential Equations—Computer/Technology-Based Texts and Courses

Web/Computer-Based Courses

C.3. Provide a broad view of the mathematical sciences

Discrete Mathematics and Data Analysis

Geometry and Geometric Thinking

Statistics and Probability and Data Analysis

Linkages—Algebra and Discrete Mathematics

Linkages—Algebra and Geometry

Linkages—Number Theory and Geometry

Linkages—Complex Variables and Geometry

Linkages—Probability and Matrix Algebra, Probability and Analysis

Powerful Applications and Contemporary Questions

Breadth of Mathematics and Connections to other Disciplines

Broader and More Flexible Major

C.4. Require study in depth

Pairs of Courses

Capstone Courses and Projects

C.5. Create interdisciplinary majors

Joint Majors

C.6. Encourage and nurture mathematical sciences majors

Prioritizing Introductory Courses

Encouraging Prospective Majors from Under-represented Groups

Providing Career Information

Mentoring and Advising Mathematics Majors

Co-Curricular Activities for Mathematics Majors

Student Assistants and Mentors

D. Mathematical sciences majors with specific career goals

D.1. Majors preparing to be secondary (9–12) school teachers

Connecting Students' Learning to their Future Teaching

Geometry

Capstone Courses for Secondary Teachers

Programs for Mathematicians Teaching Future Teachers

Collaboration with Local School Districts

D.2. Majors preparing for the non-academic workforce

Skills Needed for Industry

Advising and Mentoring for the Nonacademic Workforce

Internships and Summer Research

Professional Master's Degree

Additional Resources

D.3. Majors preparing for post-baccalaureate study in the mathematical sciences and allied disciplines

Internships and Summer Research

Special Programs for Graduate School Preparation

Mentoring and Supporting Underrepresented Students

Advising Mathematics Students

Program Examples

Annotated Bibliography

References for *CUPM Guide 2004*

American Mathematical Association of Two-Year Colleges, *Crossroads in Mathematics: Standards for Introductory College Mathematics Before Calculus*, AMATYC, 1995, `www.amatyc.org/Crossroads/CrsrdsXS.html`.

American Mathematical Society, Annual Survey of the Mathematical Sciences, `www.ams.org/employment/surveyreports.html`.

American Mathematical Society, Mathematical Association of America, and Society for Industrial and Applied Mathematics, Mathematical Sciences Careers, `www.ams.org/careers/`.

American Statistical Association/Mathematical Association of America Joint Committee on Undergraduate Statistics, Guidelines, `www.amstat.org/education/Curriculum_Guidelines.html`.

Association of American Colleges and Mathematical Association of America Joint Task Force, "Challenges for College Mathematics: An Agenda for the Next Decade, 1990, reprinted in *Heeding the Call for Change: Suggestions for Curricular Action*, L.A. Steen, ed., MAA Notes **22**, MAA, 1992.

Association of American Colleges, *Integrity in the College Curriculum*, 1985.

Association for Computing Machinery and the Computer Science Division of the Institute of Electrical and Electronics Engineers, *Computing Curriculum 2001*, `www.computer.org/education/cc2001/final/`.

Association for Women in Mathematics, `www.awm-math.org`.

Banchoff, Thomas, "Some Predictions for the Next Decade," in *CUPM Discussion Papers about Mathematics and the Mathematical Sciences in 2010: What Should Students Know?*, MAA Reports, 2001.

Clemens, Herb, "The Mathematics-Intensive Undergraduate Major," in *CUPM Discussion Papers about Mathematics and the Mathematical Sciences in 2010: What Should Students Know?*, MAA Reports, 2001.

Cobb, George, "Teaching Statistics," in *Heeding the Call for Change: Suggestions for Curricular Action*, L.A. Steen, ed., MAA Notes **22**, MAA, 1992.

COMAP, the Consortium for Mathematics and Its Applications, `www.comap.com`.

Committee on the Undergraduate Program in Mathematics, Illustrative Resources for *CUPM Guide 2004*, `www.maa.org/cupm/`.

Conference Board of the Mathematical Sciences, *The Mathematical Education of Teachers,* Issues in Mathematics Education, volume 11, AMS and MAA, 2001, `www.cbmsweb.org/`.

Dossey, John A., editor, *Confronting the Core Curriculum: Considering Change in the Undergraduate Mathematics Major*, MAA Notes **45**, MAA, 1998.

Dubinsky, Ed, David Mathews, and Barbara E. Reynolds, editors., *Readings in Cooperative Learning for Undergraduate Mathematics*, MAA, 1997.

Ellis, Wade Jr., "Mathematics and the Mathematical Sciences in 2010: What Should Graduates Know?" in *CUPM Discussion Papers about Mathematics and the Mathematical Sciences in 2010: What Should Students Know?*, MAA Reports, MAA, 2001.

Ewing, John, editor, *Towards Excellence: Leading a Doctoral Mathematics Department in the 21st Century*, American Mathematical Society Task Force on Excellence, AMS, 1999.

Ferrini-Mundy, Joan and Bradford Findell, "The Mathematical Education of Prospective Teachers of Secondary School Mathematics: Old Assumptions, New Challenges," in *CUPM Discussion Papers about Mathematics and the Mathematical Sciences in 2010: What Should Students Know?*, MAA Reports, MAA, 2001.

Ganter, Susan and William Barker, *The Curriculum Foundations Project: Voices of the Partner Disciplines*, edited and with an introduction "A Collective Vision: Voices of the Partner Disciplines," MAA, 2004.

Garfield, Joan, Bob Hogg, Candace Schau, and Dex Whittinghill, "First Courses in Statistical Science: The Status of Educational Reform Efforts," *Journal of Statistics Education*, Volume 10, Number 2, 2002. `www.amstat.org/publications/jse/v10n2/garfield.html`.

Garfunkel, Sol A., and G. S. Young, "Mathematics Outside Mathematics Departments,"*Notices of the American Mathematical Society*, **37**,1990.

———, "The Sky is Falling," in the *Notices of the American Mathematical Society*, **45,** 1998.

Gold, Bonnie, Sandra Z. Keith, and William A. Marion, editors, *Assessment Practices in Undergraduate Mathematics*, MAA, 1999, available at `www.maa.org/saum/`.

Hagelgans, Nancy L., *et al.,* editors, A *Practical Guide to Cooperative Learning in Collegiate Mathematics,* MAA, 1995.

Howe, Roger, "Two Critical Issues for the Math Curriculum," in *CUPM Discussion Papers about Mathematics and the Mathematical Sciences in 2010: What Should Students Know?*, MAA Reports, 2001.

Ingersoll, R. M., "The Problem of Underqualified Teachers in American Secondary Schools," Educational Researcher, Vol. 28, No. 2, March 1999.

Journal of Online Mathematics and its Applications (JOMA), `www.joma.org`, published by the MAA.

Keith, Sandra, "Accountability in Mathematics: Elevate the Objectives!" in *CUPM Discussion papers about Mathematics and the Mathematical Sciences in 2010: What Should Students Know?* MAA Reports, MAA 2001.

Loftsgaarden, Don O., Donald C. Rung, and Ann E. Watkins, *Statistical Abstract of Undergraduate Programs in the Mathematical Sciences in the United States, Fall 1995 CBMS Survey,* MAA Report, MAA, 1997.

Lutzer, David J., James W. Maxwell, and Steven R. Rodi, *CBMS 2000: Statistical Abstract of Undergraduate Programs in the Mathematical Sciences in the United States*, AMS, 2002.

Lutzer, David J., "Mathematics Majors 2002," *Notices of the American Mathematical Society*, Vol. 50, No. 2, February 2003.

Meier, J. and Thomas Rishel , *Writing in the Teaching and Learning of Mathematics*, MAA Notes **48**, MAA 1998.

Mathematical Association of America, *Guidelines for Programs and Departments in Undergraduate Mathematical Sciences,* MAA, 2001, available at `www.maa.org/guidelines/guidelines.html`.

Mathematical Association of America, Strengthening Under-represented Minority Mathematics Achievement, `www.maa.org/summa`.

Mathematical Association of America, Supporting Assessment in Undergraduate Mathematics (SAUM), including the MAA assessment guidelines, a volume of case studies, and syntheses of case studies on assessment at `www.maa.org/saum/`.

McCray, Patrick D., "Business View on Math in 2010 C.E.," in *CUPM Discussion Papers about Mathematics and the Mathematical Sciences in 2010: What Should Students Know?*, MAA Reports, 2001.

National Center for Education Statistics, `www.nces.ed.gov`.

National Council of Teachers of Mathematics, *Principles and Standards for School Mathematics*, NCTM, 2000, `standards.nctm.org`.

National Research Council, *BIO 2010: Transforming Undergraduate Education for Future Research Biologists*, The National Academies Presses, 2002.

National Research Council. *Educating Teachers of Science, Mathematics, and Technology: New Practices for the New Millennium*, National Academy Press, 2001.

National Science Foundation, Science and Engineering Indicators 2002, `www.nsf.gov/sbe/srs/`.

National Science Foundation , National Survey of Recent College Graduates, `www.nsf.gov/sbe/srs/`.

Resnick, L. B., "The development of mathematical intuition," in M. Perlmutter (Ed.), *Perspectives on intellectual development: The Minnesota Symposia on Child Psychology* (Vol. 19, pp. 159–194), Erlbaum, 1986.

Rogers, Elizabeth C., *et al*, editors, *Cooperative Learning in Undergraduate Mathematics*, MAA, 2001.

Sanchez, David A., "The Mathematics Major Overview," in *CUPM Discussion Papers about Mathematics and the Mathematical Sciences in 2010: What Should Students Know?,* MAA Reports, MAA, 2001.

Society for Industrial and Applied Mathematics, "SIAM Survey to Target Professional Master's Programs," *SIAM News*, Volume 36, Number 1, January/February 2003.

Steen, Lynn A., editor, *Heeding the Call for Change: Suggestions for Curricular Action*, MAA Notes **22**, MAA, 1992.

————, editor, *Mathematics and Democracy: The Case for Quantitative Literacy*, National Council on Education and the Disciplines, 2001.

————, editor, *Reshaping College Mathematics: A project of the Committee on the Undergraduate Program in Mathematics*, MAA Notes **13**, MAA, 1989.

Sterrett, Andrew, editor, *101 Careers in Mathematics*, 2nd edition, Classroom Resource Materials, MAA, 2003.

Strang, Gilbert, "Teaching and Learning on the Internet," in *CUPM Discussion Papers about Mathematics and the Mathematical Sciences in 2010: What Should Students Know?*, MAA Reports, MAA, 2001.

Tucker, Alan C., editor, *Models That Work: Case Studies in Effective Undergraduate Mathematics Programs,* MAA Notes **38**, MAA, 1995.

Wood, Susan S., "First Steps: The Role of the Two-Year College in the Preparation of Mathematics-Intensive Majors," in *CUPM Discussion Papers about Mathematics and the Mathematical Sciences in 2010: What Should Students Know?*, MAA Reports, 2001.

Wu, H. "On the Education of Mathematics Majors," in *Contemporary Issues in Mathematics Education*, ed. by E. Gavosta, S.G. Krantz and W.G. McCallum, MSRI Publications, Volume 36, Cambridge University Press, 1999, `www.math.berkeley.edu/~wu`.

Appendix 1
The CUPM Curriculum Initiative

Begun at Mathfest 1999 under the leadership of CUPM chair Tom Berger, the CUPM curriculum initiative focused on what students should know and experience as they complete their coursework in mathematics. The working assumptions of the initiative were (1) One curriculum is not appropriate for all majors; students' needs and goals have expanded, and the mathematical preparation of a diverse audience calls for a broader, more flexible major. (2) The mathematics program must serve a wide variety of mathematics-intensive majors and be responsive to the needs of other disciplines. (3) It must serve the quantitative literacy needs of a very large population often enrolled in, but not optimally served by, college algebra courses.

Gathering information

CUPM began its work on the *CUPM Guide 2004* by examining past CUPM recommendations. The 1981 recommendations recognized that many students wish to combine the study of mathematics with that of other disciplines in order to broaden their knowledge base and enhance their future career opportunities. The 1981 recommendations were reissued in 1988 in *Reshaping College Mathematics* (MAA Notes 13). The 1991 CUPM recommendations appear at the end of *Heeding the Call for Change* (MAA Notes 22). By this time, a list of courses for the major was no longer easy to state and the mathematics community was engaged in substantial discussions about calculus. Therefore, the 1991 curriculum document is brief. Since 1991, *Models That Work* (MAA Notes 38) described exemplary programs in mathematics, and *Confronting the Core Curriculum* (MAA Notes 45) addressed preparation in the first two years. The recent book on assessment, *Assessment Practices in Undergraduate Mathematics* (MAA Notes 49), examined a broad range of assessment issues and included models illustrating productive approaches. The recommendations of the Joint committee on Undergraduate Statistics of the American Statistical Association and the MAA also appear *in Heeding the Call for Change*. The recommendations of the American Mathematical Association of Two-Year Colleges appear in *Crossroads in Mathematics: Standards for Introductory College Mathematics Before Calculus*, AMATYC, 1995. All of these documents remain useful for departments planning their programs. They also serve as a base for the current report.

Since the summer of 1999, CUPM has gathered information directly from the profession in the following ways.

Meeting sessions and panels

- At Mathfest 1999, CUPM sponsored a panel/audience reaction session in front of an involved and packed audience.
- At the January 2000 Joint Meetings CUPM sponsored a panel/audience reaction session again to a packed room, plus two very well-attended contributed paper sessions.
- At Mathfest 2000 CUPM sponsored a panel of industry members commenting on the curriculum for a full and active audience.

- More panels were held at the Joint Meetings in January 2001, January 2002, and January 2003 and at August 2001.

Focus groups

- In January 2000, CUPM invited mathematicians to participate in a number of focus groups discussing curricular issues, clustering participants by institution type.
- More focus groups met in January 2001 (including groups addressing the CRAFTY Curriculum Foundations project reports—see Appendix 2), January 2002, August 2002 (including groups representing AMS and AMATYC), and January 2003.

Interdisciplinary Conferences

- See Appendix 2 on the CRAFTY Curriculum Foundations project.

Invited papers followed by a workshop discussion—September 2000

- With funding from the Calculus Consortium for Higher Education, CUPM solicited papers from a number of mathematicians asking them to address issues central to developing a planning document for departments. Writers, members of CUPM and a few others met in September 2000 for a workshop informed by and taking off from the issues raised in the papers. The invited papers appear in the January 2001 MAA Report *CUPM Discussion Papers about Mathematics and the Mathematical Sciences in 2010: What Should Students Know?* along with a summary of the deliberations at the workshop and the first tentative formulation of these recommendations. This MAA report is still available on MAA Online.

Surveys spring 2001

- Random sample of 300 departments—see Appendix 4.
- Selected sample of departments (the source of some of the examples in Illustrative Resources).

Details on the selected sample:

Questionnaires were sent to 300 selected mathematics departments. Thirty-seven were returned. Of the 37 responses, 10 indicated no data to share, 4 had some descriptive data to share. Of the remaining 23, nearly all were contacted for further information.

Forming the selected sample of 300 mathematics departments:

Thirty departments were from 2-year colleges: 9 from a list of NSF grantees, 12 from a list of FIPSE grantees, and 9 suggested by CUPM members.

The list of 270 departments at 4-year institutions was put together as follows. Working from a list of all mathematics departments provided by Jim Maxwell of the AMS, three different indices were constructed, each based on the ratio of the number of mathematics majors to the total undergraduate enrollment in 1996. The 1996 total was available for almost every department, but the numbers of majors (in 1996, 1998 and 1999) were available in different years for different institutions. The 1996 numbers were from a different survey than the 1998 and 1999 numbers and were available for more than twice as many institutions. A list was constructed of all departments ranking in the top ten percent on any one of the indices, which gave a list of 199 departments. (There were only 51 duplicates on the three lists, of 134, 56 and 60 departments calculated for numbers of majors in 1996, 1998 and 1999 respectively.)

The ratios of mathematics enrollment to total enrollment (since this is a measure of service) and of major enrollment to mathematics enrollment (a measure of recruiting success) were also examined, but these ratios could only be formed for 1998 and 1999 (and the latter measure gave a list very similar to the major to total ratio), so the ratio of majors to total undergraduate enrollment was used.

Additional departments suggested by CUPM members were added; many were already on the list. More departments were added from lists of NSF grantees and FIPSE grantees; all departments on the NSF list and all the university departments on the FIPSE were added, since relatively few universities were among those having a high major ratio. That brought the total to 300.

Involvement of partner disciplines and other professional societies

- Representatives of professional societies in engineering, physics, and economics attended the September 2000 Workshop. The presidents of MAA, SIAM, and AMATYC, and the chair of the AMS Committee on Education contributed papers and participated in the workshop.

- The AMS has representation on CUPM both through two designated AMS members on CUPM (currently Ramesh Gangolli and Diane Herrmann; previously Amy Cohen-Corwin and Naomi Fisher) and also through a liaison to the CUPM curriculum initiative (David Bressoud) appointed by the President of the AMS.

- AMATYC has a liaison to the CUPM curriculum initiative.

- The immediate past chair of the ASA/MAA Joint Committee on Undergraduate Statistics (Allan Rossman) serves on CUPM.

- The Curriculum Foundations project (Appendix 2) involved scores of representatives of mathematics-intensive disciplines.

- The CUPM curriculum initiative has an advisory committee that includes representatives of biology (Lou Gross), computer science (Peter Henderson) and engineering (David Bigio).

- Professional organizations were invited to form Association Review Groups in spring 2003 to provide comments on and reactions to draft 4.2 of the *Guide*. Reports and suggestions for improvement were received from the American Mathematical Association of Two Year Colleges, the American Mathematical Society, the American Statistical Association, the Institute of Electrical and Electronic Engineers Computer Society, the National Council of Teachers of Mathematics, and the Society for Industrial and Applied Mathematics.

Reports of other MAA committees and taskforces

In addition to CRAFTY's Curriculums Foundations project, this *Guide* has been informed by the work of the Committee on the Mathematical Education of Teachers and the report of the Conference Board of the Mathematical Sciences, *Mathematical Education of Teachers*, and reports from and information gathered by the Committee on the Teaching of Undergraduate Mathematics, the Committee on Computers in Mathematics Education, the CUPM subcommittee on Quantitative Literacy Requirements, the Committee on Articulation and Placement, and the informal working group on the first college course.

The steering committee and advisory committee

The project has been guided by a steering committee consisting of William Barker, Thomas Berger, David Bressoud (AMS representative), Susanna Epp, William Haver, Herbert Kasube and Harriet Pollatsek (chair). A larger advisory committee also includes James Lewis (chair of the MAA Coordinating Council on Education), and the members of other disciplines listed earlier. The associate directors of the MAA, first Tom Rishel and now Michael Pearson, have provided staff support.

The writing process

A writing group chaired by Harriet Pollatsek and consisting of William Barker, David Bressoud, Susanna Epp, Susan Ganter, and Bill Haver, was formed in January 2001. It developed a document for discussion at the August 2001 and January 2002 MAA meetings. In the spring and summer of 2002, with assistance from Barry Cipra, the group prepared a preliminary draft report for discussion at the August 2002 MAA meeting. Using feedback from the focus groups at that meeting, the members of the writing group began intensive work on the *CUPM Guide 2004*, which they wrote between September 2002 and August 2003. Help with assembling and editing the Illustrative Resources for CUPM Guide 2004 was provided by Kathleen Snook in the winter and spring of 2003. The writing and distribution of the *CUPM Guide 2004* was supported by grants from the National Science Foundation (DUE-0218773) and the Calculus Consortium for Higher Education.

Appendix 2
The Curriculum Foundations Project

The CUPM subcommittee *Curriculum Renewal Across the First Two Years* (CRAFTY[87]) has gathered input from partner disciplines through a series of eleven workshops held across the country from November 1999 to February 2001, followed by a final summary conference in November 2001. Each Curriculum Foundations workshop consisted of 20–35 participants, the majority chosen from the discipline under consideration, the remainder chosen from mathematics. The workshops were not intended to be discussions between mathematicians and colleagues in the partner disciplines, although this certainly happened informally. Instead, each workshop was a dialogue among the representatives from the partner discipline, with mathematicians present only to listen and serve as resources when questions arose about the mathematics curriculum.

Each workshop produced a report summarizing its recommendations and conclusions. The reports were written by representatives of the partner disciplines and directed to the mathematics community. This insured accurate reporting of the workshop discussions while also adding credibility to the recommendations. Uniformity was achieved across the reports through a common set of questions that was used to guide the discussions at each workshop (see the end of this appendix). Having these common questions also made it easier to compare the recommendations from different disciplines.

Funding for most workshops was provided by the host institutions.[88] Such financial support—obtained with little advance notice—indicates the high level of support from university administrations for such interdisciplinary discussions about the mathematics curriculum. Workshop participants from the partner disciplines were extremely grateful—and surprised—to be invited by mathematicians to state their views about the mathematics curriculum. That their opinions were considered important and would be taken seriously in the development of the *CUPM Curriculum Guide* only added to their enthusiasm for the project as well as their interest in continuing conversations with the mathematics community.

In November 2001, invited representatives from each disciplinary workshop gathered at the U.S. Military Academy in West Point, NY for a final Curriculum Foundations Conference. The discussions resulted in *A Collective Vision*, a set of commonly shared recommendations for the first two years of undergraduate mathematics instruction.

All of the Curriculum Foundations reports, along with the *Collective Vision* recommendations, have been published as an MAA Report.[89] These disciplinary reports and *Collective Vision* were heavily drawn

---

[87] Formerly Calculus Renewal And The First Two Years, but renamed when its charge widened.

[88] The only exceptions were the two workshops on technical mathematics, which were hosted by two-year institutions and funded by the National Science Foundation, and the workshop on statistics, which was mostly funded by the American Statistical Association.

[89] *The Curriculum Foundations Project: Voices of the Partner Disciplines.* Electronic versions of these materials are also available for downloading from `www.maa.org/cupm/crafty`.

upon during the construction of the *CUPM Curriculum Guide*. However, the workshop reports and *Collective Vision* have value independent of the *Guide*: they can and should serve as resources for starting multi-disciplinary discussions at individual institutions. Promoting and supporting informed interdepartmental discussions about the undergraduate curriculum might ultimately be the most important outcome of the Curriculum Foundations project.

The Curriculum Foundations Workshops

Physics and Computer Science

Bowdoin College, Maine, October, 1999
William Barker,

Interdisciplinary (Math, Physics, Engineering)

USMA, West Point, November, 1999
Don Small,

Engineering

Clemson University, May, 2000
Susan Ganter,

Health-related Life Sciences

Virginia Commonwealth University, May, 2000
William Haver,

Technical Mathematics (at two sites)

Los Angeles Pierce College, California, October, 2000
Bruce Yoshiwara,
J. Sargeant Reynolds CC, Virginia, October, 2000
Susan Wood,
Mary Ann Hovis,

Statistics

Grinnell College, October, 2000
Thomas Moore,

Business, Finance and Economics

University of Arizona, October, 2000
Deborah Hughes Hallett,
William McCallum,

Mathematics Education

Michigan State University, November, 2000
Sharon Senk,

Biology and Chemistry

Macalester College, November, 2000
David Bressoud,

Mathematics Preparation for the Major

Mathematical Sciences Research Institute, February, 2001
William McCallum,

Questions Provided at Disciplinary Workshops

Understanding and Content

- What conceptual mathematical principles must students master in the first two years?
- What mathematical problem solving skills must students master in the first two years?
- What broad mathematical topics must students master in the first two years? What priorities exist between these topics?
- What is the desired balance between theoretical understanding and computational skill? How is this balance achieved?
- What are the mathematical needs of different student populations and how can they be fulfilled?

Instructional Techniques

- What are the effects of different instructional methods in mathematics on students in your discipline?
- What instructional methods best develop the mathematical comprehension needed for your discipline?
- What guidance does educational research provide concerning mathematical training in your discipline?

Technology

- How does technology affect what mathematics should be learned in the first two years?
- What mathematical technology skills should students master in the first two years?
- What different mathematical technology skills are required of different student populations?

Instructional Interconnections

- What impact does mathematics education reform have on instruction in your discipline?
- How should education reform in your discipline affect mathematics instruction?
- How can dialogue on educational issues between your discipline and mathematics best be maintained?

Appendix 3
Data on Numbers of Majors

The declining annual number of bachelor's degrees in mathematics

As noted in the introduction, the annual number of degrees in the mathematical sciences has remained overall flat and the number of degrees in mathematics has fallen during a time when the numbers of students earning degrees in science, engineering and technology is growing. Table 3-1 shows data on the number of degrees in the mathematical sciences from the Fall 2000 CBMS Survey.[90]

Table 3-1. *Number of bachelor's degrees awarded by departments of mathematics and statistics in academic years 1979–80, 1984–85, 1989–90, 1994–95, 1999–2000.*

Academic year	79–80	84–85	89–90	94–95	99–00
Math	11,541	13,171	13,303	12,456	10,759
Math Ed	1,752	2,567	3,116	4,829	4,991
Subtotal	13,293	15,738	16,419	17,285	15,750
Stat-related*	613	659	987	1,839	1,123
Total**	13,90	19,237	19,380	20,154	19,299

*Stat-related majors include statistics, actuarial mathematics and joint mathematics/statistics majors. Of 664 statistics majors in 99–00, 394 (59%) were awarded by statistics departments; all 425 actuarial degrees in 99–00 were awarded by mathematics departments.
** The additional majors contributing to the total awarded by mathematics and statistics departments are operations research, computer science, joint mathematics/computer science, and "other" (a mixed category growing from 0 in 1979–80 to 1,507 in 1999–2000).

The data in Table 3-1 show strong growth in statistics-related degrees from 1980 until 1995, but then a 39% decline in the most recent five year period. Much of the growth in degrees awarded annually by mathematics and statistics departments was in mathematics education, which rose almost 55% from 1990 to 1995 but then remained essentially unchanged. The subtotal of mathematics and mathematics education degrees grew 30% from 1980 to 1995, but then it fell, dropping 9% in the most recent period. The annual number of degrees in mathematics grew 15 % from 1980 to 1990, but then it fell 19% in the next decade.

The serious shortage of teachers of mathematics in secondary schools makes the data on majors especially troubling. In 2000 the number of new graduates who were interested in secondary teaching fell far short of the need for mathematics teachers. The American Association for Employment in Education studied teacher supply and demand in 2000. They surveyed all institutions of higher education listed in the Higher

---

[90]*Statistical Abstract of Undergraduate Programs in the Mathematical Sciences in the United States: Fall 2000 CBMS Survey*, D.J. Lutzer, J.W. Maxwell, S.R. Rodi, AMS, 2002, Table SE 4, p. 14. Data from other sources vary somewhat. For a discussion of this variation, see "Mathematics Majors 2002," D.J. Lutzer, *Notices* of the AMS, **50** (February 2003).

Education Directory as preparing teachers. Respondents (deans and career service representatives) rated the job market for their graduates in different fields from 1 (many more applicants than jobs) to 5 (vice versa). For mathematics education, the national average was 4.44 (up from 4.18 in 1999) and varied from 3.5 in Alaska to 4.85 in the Rocky Mountain states.[91] An American Federation of Teachers survey of certified applicants to school districts in 1998–99 obtained similar results. Using the same 1–5 scale as the AAEE study, the AFT average for mathematics was 4.51.[92]

The authors of *Models that Work* point out that in the 1960s, 5% of freshmen entering colleges and universities were interested in mathematics, and 2% majored in mathematics, and this led some to the conclusion that departments were "filtering" prospective mathematics majors. Both percentages have fallen. The percentage of mathematics degrees among all bachelor's degrees was 1.54% in 1985, 1.33% in 1991, 1.17% in 1995 and 1.05% in 1998, based on NSF data.[93] For the past twenty years, the percentage of entering freshmen intending to major in mathematics has been smaller than the percentages graduating with majors in the discipline; currently the percentage is about 0.6%. Therefore, some students have been making their decision to major in mathematics during the first two years of college-level study. An optimistic interpretation of the data suggests that introductory mathematics courses have been awakening interest and encouraging students to consider majors. But departments are up against a stiff challenge, because the numbers of students entering college interested in mathematics is low, especially among groups traditionally under-represented in mathematics. Table 3-2 shows NSF figures for mathematics and, for comparison, for computer science.[94]

Table 3-2. *Percent of freshmen intending to major in mathematics/ statistics (MA) or in computer science (CS), by race and gender*

Year	1975	1980	1985	1990	1995	2000
Whites						
All (MA)	1.5	0.9	1.1	0.9	0.7	0.7
All (CS)	0.8	2.4	2.0	1.3	1.8	3.0
Females (MA)	1.5	0.9	1.1	0.8	0.7	0.6
Females (CS)	0.6	2.0	1.2	0.7	0.6	0.8
Blacks						
All (MA)	0.8	0.8	0.7	0.5	0.7	0.5
All (CS)	0.7	4.0	6.4	3.9	4.6	6.0
Females (MA)	0.7	0.7	0.6	0.5	0.7	0.5
Females (CS)	0.8	3.9	5.9	3.5	3.7	4.2
Hispanic						
All (MA)	1.8	0.8	0.7	0.8	0.5	0.6
All (CS)	1.0	2.4	2.6	1.5	2.2	2.7
Females (MA)	1.3	1.0	0.5	0.8	0.5	0.5
Females (CS)	0.6	2.8	1.7	1.2	1.1	1.1

While the percentage of White students entering college intending to major in mathematics held steady from 1995 to 2000, the percentage of females and of Blacks continued to fall. The percentages intending to major in computer science have steadily increased (except for While and Hispanic females).

---

[91] See www.ub-careers.buffalo.edu/aaee/S_DReport2000.pdf.

[92] See www.aft.org/research/survey/figures/, Figure IV-1

[93] From NSF *Science and Engineering Indicators 2002*, quoted in "Mathematics Majors 2002," D.J. Lutzer, *Notices* of the AMS.

[94] www.nsf.gov/sbe/srs Science and Engineering Indicators 2002, Appendix Table 2-11. "Hispanic" refers to Mexican American, Chicano, and Puerto Rican American.

Some of these racial and gender patterns are also visible in the number of degrees in mathematics and computer science. Tables 3-3a and 3-3b[95] show that both disciplines have improved the representation of these groups, and that women make up a much higher proportion of bachelor's degrees in mathematics than in computer science, while Blacks make up a slightly higher proportion of degrees in computer science than in mathematics. A cautionary note on these tables: some computer science degrees (about 3300 per year according to CBMS2000) are granted by mathematical sciences departments.

Table 3-3a. *Annual number (percent of total) of bachelor's degrees in mathematics*

Year	1977	1998
Total	14,303	12,363
Women	5,949 (41.6%)	5,659 (45.8%)
Blacks	712 (5.0%)	1,030 (8.3%)
Hispanic Americans	321 (2.2%)	642 (5.2%)

Table 3-3b. *Annual number (percent) of bachelor's degrees in computer science*

Year	1977	1998
All	6,426	27,674
Women	956 (14.9%)	7,439 (26.9%)
Blacks	361 (5.6%)	2,580 (9.3%)
Hispanic Americans	114 (1.8%)	1,410 (5.1%)

The annual number of women earning bachelor's degrees in mathematics fell from 1977 to 1998, but the annual number of White men earning these degrees fell even more, so the percentage of female degree recipients rose. For the same period, the annual number of computer science degrees grew more than seven-fold for women and for Blacks and more than twelve-fold for Hispanics; percentages grew as well. For Blacks and Hispanics, the number and percentage of mathematics degrees also grew, but much less dramatically. It should be added that more recent data show the annual number of majors in computer science is leveling off and that interest in computer science among entering students is declining.[96]

[95] www.nsf.gov/sbe/srs Science and Engineering Indicators 2002, Table 2-17.

[96] "Tech's Major Decline: College Students Turning Away from Bits and Bytes, E. McCarthy, *Washington Post,* August 27, 2002, p. E01; www.washingtonpost.com.

Appendix 4
Data on Student Goals, Department Practices, and Advanced Courses

The increasing diversity of student interests and goals

Mathematics majors—like all post-secondary students—are more diverse than they were even 30 years ago. The 1997 National Survey of Recent College Graduates[97] found that 34% of bachelor's degree recipients in the mathematical and related sciences also attended a two-year college. The annual number of degrees awarded by two-year colleges increased by 24% from 1985 to 2000.[98] While enrollments in remedial courses at four-year colleges and universities fell 13% during the same period,[99] they rose 58% at two-year colleges. The two-year institutions thus play a critical role in preparing students for mathematics-intensive majors, including mathematical sciences majors.

At many institutions a large fraction of the undergraduates work at part-time or sometimes even full-time jobs while they are going to school. Even with jobs, many students graduate with a substantial debt, and that fact may influence their choices of academic programs and career paths. English is a second language for many students.

Some students arrive in college and university classrooms well-prepared, highly interested in mathematics and intending doctoral study in mathematics. Other students are less intrinsically interested in mathematics and less confident of their mathematical abilities; they may choose to major in mathematics because of its applicability in other disciplines or because it offers the promise of employment opportunities. The 1997 National Survey of Recent College Graduates found that 92% of students earning bachelor's degrees in the mathematical sciences during the academic years 1994–95 and 1995–96 went directly into the workforce after graduation.[100] Data from 2002 show that approximately 3.8% of US citizen mathematics majors go on to receive doctoral degrees in the mathematical sciences within 6 years.[101]

---

[97] Every year the NSF publishes a National Survey of Recent College Graduates, sampling those who completed their degrees during the preceding two academic years. The 1997 survey was based on those who received degrees between July 1, 1994 and June 30, 1996. The survey separately tracks degrees in mathematical and related sciences and in computer and information sciences. The 1997 study was based on 26,800 bachelor's degrees in mathematical and related sciences. See www.nsf.gov/sbe/srs/nsf01337, Table B3. Table D1 on p. 92, Table D5 on p. 97.

[98] www.nces.ed.gov website, Table 247.

[99] CBMS 2000, Table SE 3.

[100] See Table D1 on p. 92 and Table D5 on p. 97.

[101] "Mathematics Majors 2002," D.J. Lutzer, *Notices of the AMS.*

The CUPM recommendations in 1981 and 1991 especially cautioned departments to design their programs to be responsive to the needs of the overwhelming majority of majors headed for the workforce rather than graduate study in mathematics.

Department practices

In 1990, a task force of the MAA and the Association of American Colleges examined the practices of mathematical sciences departments at many institutions.[102] They found a number of common features, including "a multiple track system that addressed diverse student objectives, emphasis on breadth of study in the major, and requirements for depth that help students achieve critical sophistication." The 1991 CUPM report in fact recommended exactly these practices.

In the spring of 2001, CUPM collected information from a random sample of 300 mathematics departments offering a bachelor's degree. The CUPM sample of mathematical sciences departments was stratified according to the Carnegie classification of institutions: Pub xy for public and Pri xy for private, where $x = 1$–3 denotes the highest degree offered (1 = doctorate, 2 = master's, 3 = bachelor's) and $y = 1$–4 denotes the Carnegie rank of the institution (1 is most highly rated). Thirty percent of the questionnaires sent to the random sample of departments were returned. Table 4-1 below shows the response rates for different classes of institutions.

In the responding departments, 30% of the majors earning bachelor's degrees in 2001 were either *double* majors (completing two full majors) or *joint* majors (completing a coordinated program that is less than two full majors), so linkages with other disciplines were highly significant. In fact, about two thirds of the interdisciplinary majors reported were double majors. Obviously, such cross-disciplinary choices are more difficult if the mathematics major requires a large number of courses and only double majors are possible.

Table 4-1. *Questionnaires returned from the CUPM random sample of departments, spring 2001.*

Carnegie class.		received/sent	percent
Pri	11	2/7	29%
	12	0/2	0%
	13	0/10	0%
	14	4/5	80%
	21	16/43	37%
	22	5/15	33%
	31	17/35	49%
	32	17/78	22%
Pub	11	6/14	43%
	12	2/6	33%
	13	3/6	50%
	14	0/8	0%
	21	16/54	30%
	22	1/5	20%
	31	0/1	0%
	32	2/16	13%
Total		91/305	30%

---

[102] From "Challenges for College Mathematics: An Agenda for the Next Decade, the report of the MAA-AAC Taskforce on Study in Depth, 1990," reprinted in *Heeding the Call for Change: Suggestions for Curricular Action*, MAA Notes 22, edited by Lynn Steen, p. 188.

CUPM also collected data on major requirements at 40% of the departments in the random sample.[103] Information on major requirements was obtained from web pages (plus some follow-up phone calls) for 120 departments in the random sample.[104] Because the data collection does not cover all of the random sample, the findings must be treated with caution. Note that if a department required "algebra or analysis" it was counted as specifically requiring neither. Overall 79% of the departments require abstract algebra of all of their majors and 78% require analysis. But it is striking that 100% of the highly ranked doctoral departments require algebra, and 95% require analysis. The percent requiring analysis varied more from one kind of institution to another, ranging from 100% at the highest rated public doctoral departments to 40% at the highest rated public master's level departments. See Table 4-2 below.

Table 4-2. *Percent of departments of the specified type requiring the course of all of their majors, based on 120 departments in the CUPM 2001 random sample.*

Carnegie Class. (no.)		Alg 1	Alg 2	Ana 1	Ana 2	Geo	Pro	Mod	Sem
Pub 11	(14)	100%	64%	100%	100%	21%	0%	7%	0%
Pri 11	(7)	100%	71%	86%	86%	29%	14%	0%	0%
All 11	(21)	100%	67%	95%	95%	24%	5%	5%	0%
All 12	(8)	75%	75%	88%	88%	38%	50%	13%	25%
Pub 21	(40/54)	70%	5%	40%	8%	8%	13%	8%	30%
Pub 32	(16)	69%	63%	69%	69%	19%	38%	13%	13%
Pri 31	(35)	83%	49%	89%	89%	6%	14%	3%	20%
Total	(120)	79%	41%	78%	60%	13%	18%	7%	19%

Alg = abstract algebra; Ana = analysis or advanced calculus; Geo = geometry; Prob = probability (or, in some cases a single probability/statistics course); Mod = mathematical modeling; Res = one or more of project, research, internship or senior seminar. If a department required "algebra or analysis" it was counted as requiring neither. No distinction was made between semester and quarter courses.

Advanced courses

The variation among institutions of different kinds found in the 2001 CUPM sample data mirrors the results of the 2000 CBMS survey on the percentage of departments offering certain advanced courses. See Table 4-3 below. The CBMS 2000 survey found that, with the single exception of number theory, every advanced course tracked was *less* available in 2000–2001 than in 1995–96. (CBMS did not collect data on the availability of statistics courses in 1995–96.) Their survey also found that a wider array of advanced courses was available in doctoral level departments than in master's or, especially, bachelor's degree level departments.

Percentages below 50% in the CBMS data—which occur for at least one kind of institution for every course except algebra—suggest that there may be departments not even offering these courses in alternate years, so the courses are completely unavailable to their students. Also, between 1995–96 and 2000–01, the percentage of all departments offering advanced courses in these six topics declined significantly, with

---

[103] Because of limitations of time, it was not possible to examine all 300 departments. The higher ranking departments were included in every case, which may bias the results. The omitted groups are the public and private doctoral institutions of ranks 3 and 4, the public and private master's institutions of rank 2, the private bachelor's degree institutions of rank 2, and the single rank 1 public bachelor's degree institution in the sample. Also just 40 of 54 departments were included in classification Pub 21; other categories included were complete.

[104] CUPM thanks Anna Zwahlen-Tronick and Muluwork Geremew, student workers at Colby and Mount Holyoke Colleges respectively, for their contributions to this data collection and analysis.

an 8% decline in the availability of modern algebra, and declines from 19% to 46% in the availability of analysis, geometry, modeling, operations research and combinatorics. Note that this is not a description of what is required of majors but rather of what is *available* to students.

Table 4-3. *Percent of departments offering certain advanced courses in 1995–96 and 2000–2001, by kind of department.*[105]

	in 95–96	in 00–01	in 00–01		
	all	all	PhD	MA	BA
# depts	1369	1430	187	233	1010
algebra	77%	71%	87%	88%	63%
analysis[a]	70%	56%	90%	77%	45%
geometry	69%	56%	75%	88%	46%
models[m]	35%	24%	51%	51%	13%
OR[o]	24%	13%	14%	26%	10%
combin.[c]	24%	18%	48%	24%	11%

a: advanced calculus or real analysis   m: mathematical modeling or applied mathematics
o: operations research   c: combinatorics

One might assume that diminished availability of algebra or analysis reflects a shift to various applied tracks within the major, but note the decline in availability of modeling and operations research courses. Also both algebra and analysis are recommended for prospective secondary teachers, and their availability has fallen despite the increase in the number of bachelor's degrees in mathematics education.

Not only has the availability of advanced courses declined, the enrollment in advanced courses has also declined, as Table 4-4 shows.

Table 4-4. *Fall enrollment in advanced courses in mathematics and statistics*[106]

Enrollments in 1000s	1980	1985	1990	1995	2000
Advanced mathematics	91	138	120	96	102
Upper statistics					
math depts			38	28	35
math+stat depts	43	63	52	44	45

Enrollments in both mathematics and statistics courses recovered somewhat in 2000 from the sharp declines in 1995, although not to their 1990 levels. A study in 1989 by Garfunkel and Young examined enrollments in courses in departments other than mathematics that consisted entirely or mainly of advanced mathematics.[107] (By advanced they meant having a prerequisite at or above the level of calculus.) These courses were offered not only in departments of physics, engineering, computer science, but also in economics, political science, biology, and management. Based on a sample of 425 schools they determined that each year there are over 170,000 enrollments in advanced mathematics courses being taught outside mathematics and statistics departments. It is hard to compare to the CBMS data, partly because it's full year versus fall only and partly because the definitions of advanced are not the same. As a rough estimate, this would mean that the number of student enrollments in advanced mathematical and

---

[105] CBMS 2000, Table SE.5, p. 16.

[106] CBMS 2000, Tables A1 and A2, pp. 178, 180.

[107] See "Mathematics Outside Mathematics Departments," SA Garfunkel and GS Young, *Notices* of the AMS, **37** (April 1990) for a summary of their COMAP study, and a follow-up article "The Sky is Falling," in the *Notices* **45** (February 1998).

statistical courses taught outside of mathematics and statistics departments was about 70% of student enrollments in advanced courses taught inside of mathematics and statistics departments. No follow-up of the Garfunkel and Young study has been published, but anecdotal information suggests the proportion of advanced mathematics taught outside mathematics and statistics departments in 2000 would have been at least as large.

The decline in undergraduate geometry courses

Geometric/visual thinking is an important strand in many areas of mathematics, including contemporary applied mathematics—problems rooted in geometric questions are central to the development of new technologies such as medical imaging and robotics. However, there has been a dramatic decline in undergraduate offerings in geometry; see Table 4-5 below.[108] Further, solid geometry has long been gone from the high school curriculum.

Table 4-5. *Fall enrollments in selected undergraduate courses, 1970–2000*

Enrollments in 1000's	1970	1980	1985	1990	1995	2000
Calculus level	413	590	637	647	539	570
Discrete math (intro)	na	na	14	17	16	20
Modern algebra I, II	23	10	13	12	13	11
Discrete structures	na	na	7	3	3	5
Geometry	13	4	7	8	6	6
Total adv. mathematics		91	138	120	96	102
Total mathematics		1525	1619	1621	1471	1614

Table 4-3 on the availability of advanced courses supports the point about the decline in geometry offerings: the percentage of departments offering geometry courses has fallen from 69% to 56% from 1995–96 to 2000–2001. Also, Table 4-5, on enrollments in advanced courses, shows geometry enrollments falling along with the decline in total enrollments in advanced mathematics courses from 1985 to 1995 and then remaining flat (at almost half the 1970 enrollment) from 1995 to 2000. Notice also that both total enrollments and enrollments in advanced courses rose from 1995 to 2000, but the percentage of mathematics enrollments at the advanced level fell slightly, from 6.5% to 6.3%.

---

[108] Data on undergraduate enrollment data from Table A-1 on pp. 127 ff. of *Statistical Abstract of Undergraduate Programs in the Mathematical Sciences in the United States, Fall 1995 CBMS Survey* and Table A-1 on pp. 176 ff of *2000 CBMS Survey.*

Appendix 5
Summary of Recommendations

Part I. Recommendations for departments, programs and all courses in the mathematical sciences

1. Understand the student population and evaluate courses and programs

Mathematical sciences departments should

- Understand the strengths, weaknesses, career plans, fields of study, and aspirations of the students enrolled in their courses;
- Determine the extent to which the goals of courses and programs offered are aligned with the needs of students, as well as the extent to which these goals are achieved;
- Continually strengthen courses and programs to better align with student needs, and assess the effectiveness of such efforts.

2. Develop mathematical thinking and communication skills

Every course should incorporate activities that will help all students progress in developing analytical, critical reasoning, problem-solving, and communication skills and acquiring mathematical habits of mind. More specifically, these activities should be designed to advance and measure students' progress in learning to

- State problems carefully, modify problems when necessary to make them tractable, articulate assumptions, appreciate the value of precise definition, reason logically to conclusions, and interpret results intelligently;
- Approach problem solving with a willingness to try multiple approaches, persist in the face of difficulties, assess the correctness of solutions, explore examples, pose questions, and devise and test conjectures;
- Read mathematics with understanding and communicate mathematical ideas with clarity and coherence through writing and speaking.

3. Communicate the breadth and interconnections of the mathematical sciences

Every course should strive to

- Present key ideas and concepts from a variety of perspectives;
- Employ a broad range of examples and applications to motivate and illustrate the material;
- Promote awareness of connections to other subjects (both in and out of the mathematical sciences) and strengthen each student's ability to apply the course material to these subjects;
- Introduce contemporary topics from the mathematical sciences and their applications, and enhance student perceptions of the vitality and importance of mathematics in the modern world.

4. Promote interdisciplinary cooperation

Mathematical sciences departments should encourage and support faculty collaboration with colleagues from other departments to modify and develop mathematics courses, create joint or cooperative majors, devise undergraduate research projects, and possibly team-teach courses or units within courses.

5. Use computer technology to support problem solving and to promote understanding

At every level of the curriculum, some courses should incorporate activities that will help all students progress in learning to use technology

- Appropriately and effectively as a tool for solving problems;
- As an aid to understanding mathematical ideas.

6. Provide faculty support for curricular and instructional improvement

Mathematical sciences department and institutional administrators should encourage, support and reward faculty efforts to improve the efficacy of teaching and strengthen curricula.

Part II. Additional Recommendations Concerning Specific Student Audiences

A. Students taking general education or introductory collegiate courses in the mathematical sciences

General education and introductory courses enroll almost twice as many students as all other mathematics courses combined.[109] They are especially challenging to teach because they serve students with varying preparation and abilities who often come to the courses with a history of negative experiences with mathematics. Perhaps most critical is the fact that these courses affect life-long perceptions of and attitudes toward mathematics for many students—and hence many future workers and citizens. For all these reasons these courses should be viewed as an important part of the instructional program in the mathematical sciences.

This section concerns the student audience for these entry-level courses that carry college credit.

A.1. Offer suitable courses

All students meeting general education or introductory requirements in the mathematical sciences should be enrolled in courses designed to

- Engage students in a meaningful and positive intellectual experience;
- Increase quantitative and logical reasoning abilities needed for informed citizenship and in the workplace;
- Strengthen quantitative and mathematical abilities that will be useful to students in other disciplines;
- Improve every student's ability to communicate quantitative ideas orally and in writing;
- Encourage students to take at least one additional course in the mathematical sciences.

---

[109] According to the CBMS study in the Fall of 2000, a total of 1,979,000 students were enrolled in courses it classified as "remedial" or "introductory" with course titles such as elementary algebra, college algebra, Pre-calculus, algebra and trigonometry, finite mathematics, contemporary mathematics, quantitative reasoning. The number of students enrolled in these courses is much greater than the 676,000 enrolled in calculus I, II or III, the 264,000 enrolled in elementary statistics, or the 287,000 enrolled in all other undergraduate courses in mathematics or statistics. At some institutions, calculus courses satisfy general education requirements. Although calculus courses can and should meet the goals of Recommendation A.1, such courses are not the focus of this section.

A.2. Examine the effectiveness of college algebra

Mathematical sciences departments at institutions with a college algebra requirement should

- Clarify the rationale for the requirement and consult with colleagues in disciplines requiring college algebra to determine whether this course—as currently taught—meets the needs of their students;
- Determine the aspirations and subsequent course registration patterns of students who take college algebra;
- Ensure that the course the department offers to satisfy this requirement is aligned with these findings and meets the criteria described in A.1.

A.3. Ensure the effectiveness of introductory courses

General education and introductory courses in the mathematical sciences should be designed to provide appropriate preparation for students taking subsequent courses, such as calculus, statistics, discrete mathematics, or mathematics for elementary school teachers. In particular, departments should

- Determine whether students that enroll in subsequent mathematics courses succeed in those courses and, if success rates are low, revise introductory courses to articulate more effectively with subsequent courses;
- Use advising, placement tests, or changes in general education requirements to encourage students to choose a course appropriate to their academic and career goals.

B. Students majoring in partner disciplines

Partner disciplines vary by institution but usually include the physical sciences, the life sciences, computer science, engineering, economics, business, education, and often several social sciences.[110] It is especially important that departments offer appropriate programs of study for students preparing to teach elementary and middle school mathematics. Recommendation B.4 is specifically for these prospective teachers.

B.1. Promote interdisciplinary collaboration

Mathematical sciences departments should establish ongoing collaborations with disciplines that require their majors to take one or more courses in the mathematical sciences. These collaborations should be used to

- Ensure that mathematical sciences faculty cooperate actively with faculty in partner disciplines to strengthen courses that primarily serve the needs of those disciplines;
- Determine which computational techniques should be included in courses for students in partner disciplines;
- Develop new courses to support student understanding of recent developments in partner disciplines;
- Determine appropriate uses of technology in courses for students in partner disciplines;
- Develop applications for mathematics classes and undergraduate research projects to help students transfer to their own disciplines the skills learned in mathematics courses;
- Explore the creation of joint and interdisciplinary majors.

B.2. Develop mathematical thinking and communication

Courses that primarily serve students in partner disciplines should incorporate activities designed to advance students' progress in

---

[110] See Appendix 2 for a list of the disciplines represented at the Curriculum Foundations workshops.

- Creating, solving, and interpreting basic mathematical models;
- Making sound arguments based on mathematical reasoning and/or careful analysis of data;
- Effectively communicating the substance and meaning of mathematical problems and solutions.

B.3. Critically examine course prerequisites

Mathematical topics and courses should be offered with as few prerequisites as feasible so that they are accessible to students majoring in other disciplines or who have not yet chosen majors. This may require modifying existing courses or creating new ones. In particular,

- Some courses in statistics and discrete mathematics should be offered without a calculus prerequisite;
- Three-dimensional topics should be included in first-year courses;
- Prerequisites other than calculus should be considered for intermediate and advanced non-calculus-based mathematics courses.

B.4. Pre-service elementary (K–4) and middle school (5–8) teachers

Mathematical sciences departments should create programs of study for pre-service elementary and middle school teachers that help students develop

- A solid knowledge—at a level above the highest grade certified—of the following mathematical topics: number and operations, algebra and functions, geometry and measurement, data analysis and statistics and probability;
- Mathematical thinking and communication skills, including knowledge of a broad range of explanations and examples, good logical and quantitative reasoning skills, and facility in separating and reconnecting the component parts of concepts and methods;
- An understanding of and experience with the uses of mathematics in a variety of areas;
- The knowledge, confidence, and motivation to pursue career-long professional mathematical growth.

C. Students majoring in the mathematical sciences

The recommendations in this section refer to all major programs in the mathematical sciences, including programs in mathematics, applied mathematics, and various tracks within the mathematical sciences such as operations research or statistics. Also included are programs designed for prospective mathematics teachers, whether they are "mathematics" or "mathematics education" programs, although requirements in education are not specified in this section.

Although these recommendations do not specifically address minors in the mathematical sciences, departments should be alert to opportunities to meet student needs by creating minor programs—for example, for students preparing to teach mathematics in the middle grades.

These recommendations also provide a basis for discussion with colleagues in other departments about possible joint majors with any of the physical, life, social or applied sciences.

C.1. Develop mathematical thinking and communication skills

Courses designed for mathematical sciences majors should ensure that students

- Progress from a procedural/computational understanding of mathematics to a broad understanding encompassing logical reasoning, generalization, abstraction and formal proof;
- Gain experience in careful analysis of data;
- Become skilled at conveying their mathematical knowledge in a variety of settings, both orally and in writing.

C.2. Develop skill with a variety of technological tools

All majors should have experiences with a variety of technological tools, such as computer algebra systems, visualization software, statistical packages, and computer programming languages.

C.3. Provide a broad view of the mathematical sciences

All majors should have significant experience working with ideas representing the breadth of the mathematical sciences. In particular, students should see a number of contrasting but complementary points of view:

- Continuous and discrete,
- Algebraic and geometric,
- Deterministic and stochastic,
- Theoretical and applied.

Majors should understand that mathematics is an engaging field, rich in beauty, with powerful applications to other subjects, and contemporary open questions.

C.4. Require study in depth

All majors should be required to

- Study a single area in depth, drawing on ideas and tools from previous coursework and making connections, by completing two related courses or a year-long sequence at the upper level;
- Work on a senior-level project that requires them to analyze and create mathematical arguments and leads to a written and an oral report.

C.5. Create interdisciplinary majors

Mathematicians should collaborate with colleagues in other disciplines to create tracks within the major or joint majors that cross disciplinary lines.

C.6. Encourage and nurture mathematical sciences majors

In order to recruit and retain majors and minors, mathematical sciences departments should

- Put a high priority on effective and engaging teaching in introductory courses;
- Seek out prospective majors and encourage them to consider majoring in the mathematical sciences;
- Inform students about the careers open to mathematical sciences majors;
- Set up mentoring programs for current and potential majors, and offer training and support for any undergraduates working as tutors or graders;
- Assign every major a faculty advisor and ensure that advisors take an active role in meeting regularly with their advisees;
- Create a welcoming atmosphere and offer a co-curricular program of activities to encourage and support student interest in mathematics, including providing an informal space for majors to gather.

D. Mathematical sciences majors with specific career goals

D.1. Majors preparing to be secondary school (9–12) teachers

In addition to acquiring the skills developed in programs for K–8 teachers, mathematical sciences majors preparing to teach secondary mathematics should

- Learn to make appropriate connections between the advanced mathematics they are learning and the secondary mathematics they will be teaching. They should be helped to reach this understanding in courses throughout the curriculum and through a senior-level experience that makes these connections explicit.

- Fulfill the requirements for a mathematics major by including topics from abstract algebra and number theory, analysis (advanced calculus or real analysis), discrete mathematics, geometry, and statistics and probability with an emphasis on data analysis;
- Learn about the history of mathematics and its applications, including recent work;
- Experience many forms of mathematical modeling and a variety of technological tools, including graphing calculators and geometry software.

D.2. Majors preparing for the nonacademic workforce

In addition to the general recommendations for majors, programs for students preparing to enter the nonacademic workforce should include

- A programming course, at least one data-oriented statistics course past the introductory level, and coursework in an appropriate cognate area; and
- A project involving contemporary applications of mathematics or an internship in a related work area.

D.3. Majors preparing for post-baccalaureate study in the mathematical sciences and allied disciplines

Mathematical sciences departments should ensure that

- A core set of faculty members are familiar with the master's, doctoral and professional programs open to mathematical sciences majors, the employment opportunities to which they can lead, and the realities of preparing for them;
- Majors intending to pursue doctoral work in the mathematical sciences are aware of the advanced mathematics courses and the degree of mastery of this mathematics that will be required for admission to universities to which they might apply. Departments that cannot provide this coursework or prepare their students for this degree of mastery should direct students to programs that can supplement their own offerings.

Appendix 6
Sample Questions for Department Self-Study

These sample questions are meant to be suggestive. Every department should draft its own set of questions to gauge its progress in meeting the CUPM recommendations.

1. Do we have data on subsequent course taking in mathematics by students enrolled in our introductory courses? (For example, do we know how many of our Pre-calculus students successfully complete Calculus I? How many of our Calculus I students successfully complete a second course in the department?)

2. Do we know the intended majors of the students enrolled in our introductory courses?

3. Do we know how many of our majors enter the job market directly after graduation, and what kinds of jobs they take?

4. In the past five years, have we asked our majors who graduated recently what they think of the quality of their undergraduate preparation in mathematics?

5. In the past year, has at least one member of our department had a conversation with someone from a nearby business or industry to discuss the mathematics they need in their job and the skills they look for when hiring?

6. Does every course we offer have examinations or assignments *that affect a student's grade* requiring students to
   a. explain their reasoning?
   b. solve multi-step problems?
   c. generalize from examples?
   d. solve a problem two different ways?
   e. read new material and use it in some way?
   f. use a mathematical tool to solve a problem in another discipline?

7. Does the syllabus of every course we offer include
   a. an application to at least one discipline outside mathematics?
   b. at least one topic or application that is less than fifty years old?

8. Have we used *The Curriculum Foundations Project: Voices of the Partner Disciplines* to initiate and support conversations with faculty in other disciplines?

9. In the past year, has at least one member of our department had a conversation with a faculty member from another discipline* about
   a. a course we offer that their students take?
   b. a course we might offer that would be valuable for their students?
   c. applications to their field that we might include in a course we teach?
   d. possible undergraduate research projects?

e. team-teaching (or guest lectures in) a course or a unit within a course?
(*specifically with someone from the biological sciences? business or economics? chemistry? computer science? engineering? physics?)

10. Have we had a conversation with another department about creation of a joint major?

11. Do we offer at least one introductory course in which an examination or assignment *that affects a student's grade* requires students
a. to use computer technology to solve a problem?
b. to use computer technology to investigate examples or visualize a concept?

12. Same as (11) for intermediate courses.

13. Same as (11) for advanced courses.

14. Are faculty in our department rewarded for extra teaching effort (such as learning substantial new material, extensive consultation with colleagues outside the department, or taking leadership of the curriculum and teaching of a multi-section introductory course) by one or more of the following?
* Released time.
* Credit toward merit pay, promotion or tenure.
* Travel money for professional development.
* Institutional recognition (teaching awards etc.).

15. Are faculty in our department offered support in using new technology or in learning new pedagogical strategies by one or more of the following?
* In-house workshops.
* Support to attend workshops/minicourses off campus.
* Released time.
* Extra student assistants.

16. Do we offer at least one introductory course that satisfies Recommendation A.1?

17. Do we make effective use of advising, placement tests and/or consultation with colleagues in other disciplines to ensure that students take appropriate introductory courses?

18. Do we offer a statistics course with an emphasis on data analysis and without a calculus prerequisite?

19. Do we offer a discrete mathematics course without a calculus prerequisite that meets the needs of computer science majors?

20. Do we incorporate geometric thinking and visualization in two and three dimensions — including vectors — in our first-year courses? In our second-year courses?

21. Have we examined the prerequisites for our intermediate and advanced courses with an eye to making them more accessible to students majoring in other disciplines or not yet decided on majors?

22. Have we consulted with colleagues in education about our programs for prospective teachers?

23. Do we offer a program for prospective elementary and middle school teachers that satisfies Recommendation B.4?

24. Are school systems that hire our graduates satisfied with their performance?

25. Can we see progress in our majors' abilities to reason, solve problems and think abstractly as they move through our program? How do we gauge their progress?

26. Can we see progress in our majors' abilities to read and write mathematics and present their ideas orally as they move through our program? How do we gauge their progress?

27. Do our majors have experience with a variety of technological tools? Which courses provide these experiences?

28. Does every major complete a set of courses that encompasses the breadth specified in Recommendation C.3? Which courses include which aspects of the contrasting but complementary elements listed in C.3?

29. By graduation, does every major know several substantial applications of mathematics and a number of contemporary open questions?

30. Do we assure that every major studies a single area in depth as specified in Recommendation C.4? What are the ways a student can satisfy this requirement?

31. Does every major complete a senior year project as specified in Recommendation C.4? What are the ways a student can satisfy this requirement?

32. Is our major flexible and adapted to connections to other disciplines? How do we know?

33. How do we recruit and retain majors and minors? Are we satisfied with our efforts? Are our students satisfied with our efforts?

34. How do we advise current and prospective majors? Are we satisfied with our efforts? Are our students satisfied with our efforts?

35. Do we effectively inform students about career opportunities for majors in the mathematical sciences? About what our alumni do after graduation?

36. Do courses and senior year projects for prospective teachers of secondary mathematics make explicit connections between new material and the material of the high school curriculum? What specific elements help students make these connections?

37. Does our program for prospective teachers of secondary mathematics include the topics listed in Recommendation D.1?

38. Do all prospective teachers of secondary mathematics experience many forms of mathematical modeling and a variety of technological tools? Which courses provide these experiences?

39. Do our majors preparing for the nonacademic workforce complete courses in programming, statistics and a cognate area as recommended in D.3? Which specific courses do we recommend to our students to meet these requirements?

40. Does every major preparing for the nonacademic workforce complete an internship or a project involving a contemporary application of mathematics or statistics?

41. Are employers who hire our graduates satisfied with their mathematical preparation?

42. How do we inform students about special opportunities like internships, summer research programs and visiting programs at other universities?

43. Which faculty have responsibility for advising students about post-baccalaureate study? Are students satisfied with the guidance they're getting? Are the programs that admit our students satisfied with their preparation?

44. Which faculty have responsibility for advising students intending doctoral study in the mathematical sciences? Have they contacted the graduate programs to which our students apply? Are the programs sthat admit our students satisfied with their preparation?

Index

AAC Association of American Colleges

AAEE American Association for Employment in Education

Abstract algebra 20, 38, 45, 46, 47, 49, 52, 54, 60

Accountability 12

ACM Association for Computing Machinery

Advanced courses in mathematics 21, 23, 38, 48–49, 53–55, 60, 87–88

Advising 32, 50, 51–52, 58, 59

AFT American Federation of Teachers

Algebra 38, 39, 41, 47

Algorithms 24, 55

Alumni, information from 51

AMATYC American Mathematical Association of Two-Year Colleges

AMS American Mathematical Society

Analysis 45, 46, 52, 54, 60
    Complex 47, 49, 60
    Real 49, 52,

Analytical thinking: see Reasoning and Problem solving

Applications 17, 18, 19, 20, 32, 34, 35, 36, 38, 42, 46, 48, 52, 55, 56, 57, 58

Approximation 18

Articulation 9, 30, 40, 51

ASA American Statistical Association

Assessment 11–13, 30, 31, 97–99

Banchoff, T. 23

Bigio, D. 75

Biology 19, 33, 36, 42, 51, 59, 60, 75
    Mathematical 36

Breadth, mathematical 46–48

Bridge courses: see Transition courses

Business and management 37, 42, 59

Calculus 14, 18, 19, 27, 33, 38, 45, 46, 47, 50, 54, 55, 59
    Multivariable 34, 38, 47, 49

Capstone course 49

Careers for mathematical sciences majors 50, 51, 58
    Nonacademic workforce 57–58
    Research and post-secondary teaching 58–61
    Secondary teaching 52–56

CBMS Conference Board of the Mathematical Sciences

Chaos theory 21

Chemistry 34, 38, 59

Clemens, H. 47,

Co-curricular activities for mathematics majors 50, 52

Coding theory 20, 21, 47, 49

College algebra 27, 30–31, 39

COMAP Consortium for Mathematics and its Applications

Combinatorics 21, 38, 49, 55

Communicating mathematics 13, 16, 28, 29, 37, 38, 42, 44, 45, 49, 58
    Reading 60
    Writing 17, 33 (and see Communicating mathematics)

Complex analysis: see Analysis, Complex

Complexity theory 21

Computational algebraic geometry 21, 46, 48, 49

Computational geometry 48

Computational techniques 32, 35

Computer graphics 21, 36

Computer science 35, 36, 37, 38, 75

Computer technology: see Technology

Conjecture 13, 15, 55, 56

Connections within mathematics and across disciplines 17, 20, 48, 52, 53–54

Contemporary topics 17, 20, 21, 46, 52, 55, 58

Continuous mathematics 46, 47,

Control theory 48

CRAFTY Curriculum Renewal Across the First Two Years (MAA)

Cryptography 19, 21, 28, 45, 59

CUPM Committee on the Undergraduate Program in Mathematics (MAA)

Curriculum Foundations project 77-79

Data on
    Advanced courses 4, 87–89
    Careers 57, 81–82, 83
    Degrees, bachelor's 4
    Department practices 86–89
    Enrollment 4, 27, 40, 47, 74, 88
    Diversity 85–86
    Gender 82–83
    Intended majors 44
    Majors 4, 74, 81-83
    Major requirements 87
    Race and Ethnicity 81–83
    School completion 11
    Teacher preparation 40, 43, 81–82

Data analysis 28, 29, 35, 37, 38, 41, 44, 45, 47, 52, 55, 57, 58

Depth, mathematical 48–49

Developmental mathematics 9

Difference equations 56

Differential equations 23, 24, 45, 46, 48, 55, 56, 59

Discrete mathematics 14, 27, 31, 33, 34, 35, 37, 38, 45, 46, 48, 50, 52, 55

Doctoral programs
    Allied disciplines 58
    Mathematical sciences 43, 58, 59–61
Dubinsky, E. 16
Dynamical systems 21

Economics 19, 36, 37, 38, 42, 60
Elementary school teachers: *see* Prospective teachers
Ellis, W. 23
Employers, information from 51
Engineering 37, 42, 59, 75
Euclidean geometry: *see* Geometry, Euclidean
Euclidean spaces 54
Exact solutions 18
Examples, 17, 18, 19
Exploration 13, 15, 24, 55

Faculty
    Adjunct 25
    Graduate students 25
    Part-time 25
    Support for 25–26, 34
Ferrini-Mundy, J. 53
Financial mathematics 36, 59
Findell, B. 53
Finite mathematics 28
FIPSE Fund for the Improvement of Post-Secondary Education
Fractals 21, 28
Functions 38, 41, 54, 55, 56

Game theory 21, 48, 56
Garfield, J. 31
Garfunkel, S. 88
General educations courses in mathematics 27–32
Generalization 15
Geometry 20, 38, 45, 47, 50, 52, 55, 56, 60, 89
    Algebraic 21, 46, 48, 49
    Euclidean 54, 55
    Projective 49
Geometric reasoning: *see* Reasoning, Geometric
Geremew, M. 87
Gold, B. 13
Graph theory 49, 55, 56
Gross, L. 75
Group theory 20

Hagelgans, N. 16
Henderson, P. 75
History of mathematics and its applications 52, 56
Hogg, R. 31
Hovis, M.A. 78
Howe, R. 47
Hughes Hallett, D. 78

IEEE Institute of Electrical and Electronics Engineers
Illustrative Resources 63–67
Image compression 45, 48
Information theory 48
Ingersoll, R. 43
Interdisciplinary
    Collaboration 21, 22, 32, 33, 36

Courses 21, 35, 36
    Graduate programs 59
Internships 49, 57, 58
Introductory mathematics courses 14, 27–32
    Effective teaching in 50

JOMA *Journal of Online Mathematics*

Keith, S. 12
Knot theory 21

Law schools 59
Liberal arts mathematics 28
Life sciences 42
Linear algebra 18, 19, 23, 24, 33, 34, 38, 46, 47, 49, 54, 59
Linear programming 56
Linear statistical models 47, 56
Lutzer, D. 4, 81, 85

MAA Mathematical Association of America
Majors
    Joint and interdisciplinary 21, 32, 36–37, 42, 43, 46, 49, 51
    Mathematical sciences 14, 32, 35, 42–61,
    Mathematics education 42, 52–56
    Recruiting 50–52
    Tracks in the mathematics majors 49
Management: *see* Business and management
Master's degree programs 51, 58
Mathematics 3
Mathews, D. 16
Maxwell, J. 4, 74, 81
McCallum, W. 78
McCray, P. 57
Measurement 38, 41, 56
Medicine 42, 59
Mentoring
    Faculty 25
    Students 50, 51
MET Mathematical Education of Teachers
Middle school teachers, prospective: *see* Prospective teachers
Minors in mathematics 40, 42, 51
Modeling, mathematical 15, 19, 23, 28, 29, 33, 34, 37, 45, 52, 55, 56, 58
    Deterministic 46, 47, 56
    Stochastic 56, 57
Monte Carlo methods 21, 48
Moore, T. 78
Multivariable calculus: *see* Calculus, Multivariable

NAEP National Assessment of Educational Progress
NCTM National Council of Teachers of Mathematics
NSF National Science Foundation
Number theory 21, 38, 46, 47, 50, 52, 54
Numerical analysis 49, 60

Operations research 42, 59
Optimization 48

Partner disciplines 8, 32, 75
    Courses serving 22, 32, 34, 35, 36, 37
Perspectives 17

Physical sciences 42, 60
Placement tests 12, 32
Pólya, G. 19
Post-baccalaureate study
    Allied disciplines 58–61
    Mathematical sciences 58–61
    Mathematical sciences, doctoral: *see* Doctoral programs
Pre-calculus 14, 31, 38
Prerequisites 34, 37–38, 46, 49
Probability 33, 38, 41, 46, 47–48, 49, 52, 55, 57, 58
Problem solving 13, 15, 22, 24, 47, 55, 57, 58
Professional development 25, 38, 42
Professional schools 43, 59
Professional societies 75
Programming, computer 24, 45, 46, 55, 57, 59
Project, senior-level 48, 49, 54, 56, 58
Proof 14, 24, 35, 37, 45, 55
Prospective teachers 33, 38–42
    Elementary 27, 38, 40–41
    Foundational knowledge 38, 39, 40, 41
    Middle school 27, 38, 40, 41, 42
    Secondary mathematics 42, 52–56

Quantitative literacy: 28, 29
Quantitative reasoning: *see* Reasoning, Quantitative
Quantum computing 21

Reading mathematics: *see* Communication
Reasoning 13, 14, 15, 28, 29, 37, 38, 42, 44, 45, 57
    Geometric 34, 38, 47
    Quantitative 28, 29, 38
Recruiting majors: *see* Majors, Recruiting
Research, undergraduate: *see* Undergraduate research
Resnick, l. 40
Reynolds, B. 16
Rodi, S. 4, 81
Rogers, E. 16

Sánchez, D. 39, 53
SAUM Supporting Assessment in Undergraduate Mathematics (MAA)
Schau, C. 31
Service courses: *see* Partner disciplines, Courses serving
SIAM Society for Industrial and Applied Mathematics
Small, D. 78
Speaking mathematics, *see* Communication
Statistics 14, 18, 21, 24, 27, 28, 31, 33, 34, 35, 37, 38, 41, 42, 45, 46, 47–48, 49, 50, 52, 55, 56, 58, 59, 75

Steen, L. 29
Sterrett, A. 51
Strang, G. 23, 44
Student lounge 52
Survey courses in mathematics 50
Symbol manipulation 18

Team-teaching 21
Technology, computer 21–24, 32, 35, 36, 45–46, 52, 55, 56, 58
    Computer algebra system 23, 46
    Differential equation solver 23
    Geometry software 52, 56
    Graphing calculator 36, 46, 52, 56
    Graphing utility 23,
    Spreadsheets 23, 24, 36, 46, 56
    Statistical packages 24, 45, 46
    Tutorial software 23
    Visualization software 45, 46
Theoretical mathematics 18, 46, 48
Thinking, analytical: *see* Reasoning
Thinking, mathematical: *see* Reasoning
Three-dimensional topics 34, 37, 38, 56
Topology 60 (and *see* Knot theory)
Transfer problem 20, 32, 36
Transition courses 14, 23
Tucker, A.C., 4
Tutors, undergraduate 50, 51
Two-year colleges 11, 40, 43, 50, 73, 74, 75

Undergraduate research 21, 32, 36, 49, 60
Underrepresented groups in mathematics 50, 51
Understanding mathematics 22, 24, 34, 45, 46, 52, 54, 56

Vectors 34, 38
Visualization skills 15, 24, 34, 38, 41, 47, 55

Wavelets 21, 48,
Whittinghill, D. 31
Wood, S. 11, 51, 78
Writing mathematics: *see* Communication
Wu, H. 53, 54, 55

Yoshiwara, B. 78
Young, G. 88

Zwahlen-Tronick, A. 87